REFRAMING DISEASE CONTEXTUALLY

Philosophy and Medicine

VOLUME 81

The titles published in this series are listed at the end of this volume

REFRAMING DISEASE CONTEXTUALLY

by

MARY ANN G. CUTTER

University of Colorado,
Colorado Springs, Colorado, U.S.A.

KLUWER ACADEMIC PUBLISHERS

DORDRECHT / BOSTON / LONDON

A C.I.P. Catalogue record for this book is available from the Library of Congress.

ISBN 1-4020-1796-0

Published by Kluwer Academic Publishers,
P.O. Box 17, 3300 AA Dordrecht, The Netherlands.

Sold and distributed in North, Central and South America
by Kluwer Academic Publishers,
101 Philip Drive, Norwell, MA 02061, U.S.A.

In all other countries, sold and distributed
by Kluwer Academic Publishers, Distribution Center,
P.O. Box 322, 3300 AH Dordrecht, The Netherlands.

Printed on acid-free paper

Printed in the Netherlands.

TABLE OF CONTENTS

PREFACE

This book is about disease. It is an attempt to come to terms with how we understand and undertake disease in contemporary medicine and culture. It argues that disease is contextual. More specifically, our understanding of disease is contingent on three kinds of commitments: metaphysical, epistemological, and axiological. Corresponding to the three kinds of commitments are three dimensions of clinical explanation. Disease necessarily involves all three. The interplay among these dimensions is such that disease is theory and value-laden in a way that is context-dependent. To defend this thesis, I argue that various philosophical accounts of disease differ depending on the extent to which they explicitly or implicitly hold that the character of disease is (1) a universal thing as opposed to a particular process, (2) discovered or created, and (3) value-neutral or value-ladened, and (4) relative or context-dependent. The first two can be seen as involving traditional metaphysical (Chapter 3) and epistemological (Chapters 4 and 5) controversies in the history of medicine regarding the extent to which knowledge describes structures in reality independent of the categories of the knower. The third can be seen as involving a more recent debate in the philosophy of medicine regarding the axiological character of disease (Chapters 6 and 7). The fourth restates the first three but in terms of the extent to which the theoretical and value assumptions are dependent on the context (Chapter 8). Disease is, in short, multi-dimensional. Thus, the thesis is not just that disease is best understood in terms of this four-fold analysis, but that disease is theoretical and value-ladened in a way that is context-dependent. Disease is theory-ladened in that clinical reality is seen within frameworks of reality and knowledge, value-ladened in the sense that values are inseparably bound to clinical facts, and contextual in that socio-cultural influences condition clinical theory and values. An implication of this thesis is that a universal account of disease is not forthcoming. Disease is localized or nested in particular contexts.

The terrain of the debates that is offered acquaints readers with the rich discussion of disease in the history of modern Western medicine as well as provides a background for understanding the development of accounts of disease and of specific disease types, with specific attention to major paradigms and their shifts in medicine. In this way, this book is for the individual wishing to become acquainted with the philosophy of disease. In the end, though, the account of disease offered here *reframes* earlier notions within the interacting contexts of understanding, treatment, and social forces. Theory and practice, and facts and values, interplay in such a way as to produce accounts of patient dysfunction and suffering. Numerous applications of this account are offered, as in seen in the case of AIDS (Chapter 2), genetic disease

(Chapter 9), and feminist accounts of disease (Chapter 10), among others. In this way, this book forwards a position that is worthy of consideration by current scholars.

This inquiry is the product of research and teaching at the University of Colorado, Baylor College of Medicine, and Georgetown University. I became interested in disease during undergraduate studies at Georgetown University, while in a class entitled "Facts and Explanation in Medicine." During the late 1970s, discussions emphasized the implications of labels and stereotypes in medicine and my interest in disease reflected these concerns. My first attempt at investigating disease focused on explanation in medicine. Much of my earlier reflections at Baylor College of Medicine had to do with the logic of disease. While this focus is certainly important, other issues increasingly commanded my attention, such as how disease tells us about ourselves as knowers, actors, valuers, and members of a culture. As a result, this inquiry took shape at the University of Colorado.

Some recognitions are in order. Colleagues at the University of Colorado, Raphael Sassower and Fred Bender, have each in their own distinctive and thoughtful way influenced this work. Special recognition goes to Fred Bender, who patiently read through the entire project, providing wise and rigorous insight along the way. Thanks, Fred, for those educational runs. I am indebted as well to Fred Abrams, Lynda Fox (now deceased), George Khushf, Larry McCullough, Joe McInerney, George Schroeder, and my friends at National Scoring Services for their support during my years at the University of Colorado. Their friendship and encouragement continue to be deeply appreciated. Between 1983 and 1988, I had the honor and privilege of working with scholars at the Center for Ethics, Medicine, and Public Issues (currently known as the Center for Medical Ethics and Health Policy), Baylor College of Medicine, Houston, Texas. I am particularly indebted to H. Tristram Engelhardt, Jr for his guidance and support of my work in philosophy of medicine and his introduction to the world of academic scholarship. If not for Dr. Engelhardt's invitation, I would not have left Washington, D.C. for Texas. If not for his encouragement, I would not be writing on concepts of disease. Thanks as well to Tom L. Beauchamp, Edmund D. Pellegrino, and Robert M. Veatch, all of whom mentored me during graduate work at Georgetown University. I continue to appreciate their guidance and encouragement well past my studies at Georgetown University. Finally, special thanks to the anonymous reviewers who submitted insightful and helpful comments for the revision of this project.

Others not so directly tied to academic communities are to be thanked. My father, John P. Gardell, and Vera Silvestro Gardell have provided the encouragement an offspring cherishes when pursuing important goals. My mother, M. Theresa Swords Gardell, has remained a symbol of fortitude and wisdom in my life, well after her untimely death in 1985. This book is dedicated to her. Laura Sansone Keyes and Anne Sarno have been steadfast friends throughout the years. Everyone should have such great listeners in their lives. My husband, Lewis M. Cutter, Jr., has made possible a life filled with children, teaching, writing, and climbing mountains. His non-conditional support during these years of academic toil and raising children is truly a gift. Words fail to account for the possibilities in human relationship. Our children, Lewis Moore, Theresa Morrell, and John Calvin, are our greatest treasures. They are to be thanked for calling my attention to the important things in life during my times at the desk, in the books, and on the computer. Such interruptions remind me that the concepts, ideas, and values that philosophers study are rarely so tidy, simple, and removed from the lives we live--and rarely are they more important.

University of Colorado
Colorado Springs, Colorado
August 2003

xii

LIST OF FIGURES

CHAPTER 1

INTRODUCTION

This work examines disease (ME *disese*, Ofr *desaise*, disease; *des-* priv., + *aise*, ease; L. *dis-* priv., and *esse*, to be). [1] The central thesis of this analysis is that our understanding of disease is contingent on commitments[2] to three types of presuppositions: metaphysical (which establishes the nature of reality), epistemological (which establishes how we know what we know), and axiological (which establishes what and how we value). The interplay among these three types of commitments makes disease both theory-laden and evaluative. This means that it is also best understood as context-dependent. Put another way, the term "disease" refers to descriptions of pathological processes abstracted from individual physiological processes that afflict individuals in the human species, bring patients to the attention of health care professionals, and serve as treatment warrants. Given this multi-dimensional character of disease, we can expect different understandings and responses to particular diseases.

1. WHY STUDY DISEASE?

It may come as a surprise to some that someone would question the meaning of disease. Large clinical textbooks are written on disease, such as heart disease, acquired immunodeficiency syndrome, and cancer. Vast amounts of time and money have gone into exploring diseases, and hospitals and clinics are full of their consequences. Debating the topic at all may appear to be either worthless or destructive to medicine. Yet, to adopt this attitude is to miss some critical lessons.

Decisions about the meaning of disease have direct and important consequences for daily life and the allocation of significant portions of social resources. As Arthur L. Caplan (1993) notes, the emergent concern with disease in the twentieth century is a function of many forces. To begin with, disease has served as a major classification in medicine since its beginning and it is important to be clear about what medicine is talking about when it uses the term "disease." The efficacy of twentieth century medicine, public health, and its attendant technology in preventing and reversing many forms of infection, dysfunction, and nutritional deficiency constitute reasons for attention to disease. Reproductive technologies, genetic engineering, cosmetic surgery, and physician-assisted suicide, for instance, challenge our view about what constitutes disease and the proper domains of medical attention and why. Concerns regarding the role medicine as a powerful social institution plays in assessing the worth and value of human beings through its clinical classifications are

1

additional reasons for interest in disease. One is reminded of the abuse in medicine against women (Smith-Rosenberg and Rosenberg, 1981), blacks (Cartwright, 1981 [1851]), Jews (Proctor, 1988), and Russian dissidents (Pope, 1996). Finally, and as a result of numerous forces such as legal restrictions (e.g., Colorado Revised Statute 10-3-1104.7,[3] Equal Employment Opportunity Commission, 1995[4]), institutions such as the government, insurance industry, and business increasingly play roles in determining what clinical conditions are worthy of recognition as disease by deciding which are covered and which are not. Depending on how disease is understood, the investment of resources (e.g., funds, personnel, power, legal protection) will be seen to be indicated or unnecessary.[5] Given that developed countries are spending more and more in the treatment and prevention of disease [6], the allocation of health care presents the opportunity to consider just distribution of resources.

In addition to practical reasons to attend to the nature and scope of disease, there are conceptual ones. A literature review illustrates varying definitions of disease. A major focus of discussion involves the extent to which disease is, simply put, a biological concept. On the one hand, there are those who hold that disease is reducible to biological dysfunction or disadvantage. Christopher Boorse appeals to biological advantage or statistical normality when he defines disease as "a type of internal state which is either an impairment of normal functional abilities below typical efficiency, or a limitation on functional abilities caused by environmental agents" (1977, p. 555). Similarly for J.A. Barondess, "[d]isease may be viewed as a biological event....It is a disruption in the structure and/or function of a body part or system" (1979, p. 376). E.J.M. Campbell et al. (1979) provide this definition: "disease refers to the sum of abnormal phenomena displayed by a group of living organisms...by which they differ from the norm for their species in such a way to place them at a biological disadvantage" (1979, p. 761). Feinstein states that disease is deleterious changes, autochthonous or allochthonous--that is arising from the body tissues themselves or from foreign sources; or alternatively, from breakdown of the homeostatic mechanisms" (1967, p. 120). Contemporary genetics tempts many to see disease as simply a function of abnormal molecular processes (Lewontin, 1991). In short, a prominent view in late twentieth century thinking is that disease is a biological concept and subject to (strictly speaking) scientific explanation.

In contrast, there are those who hold that the concept of disease is not reducible[7] to biological dysfunction. On this view, disease is more appropriately understood as a state constituting some harm or threat to a person's well-being, where harm or threat are terms that require non-scientific interpretation. Consider Lester S. King: "Biological science does not try to distinguish between health and disease. Biology is concerned with the interaction between living organisms and their environment. What we call health or disease is quite irrelevant" (1981 [1954], p. 107). Engelhardt indicates the importance of the nonbiological perspective when he says: "The concept of disease is used in accounting for physiological and psychological (or behavioral) disorders, offering generalizations concerning patterns of phenomena which we find disturbing and unpleasant" (1981 [1975], p. 32). Margolis says, "disease is whatever is judged to disorder or to cause disorder, in the relevant way, the minimal integrity of the body and mind relative to prudential functions" (1976, p. 253). Edmund D. Pellegrino and David C. Thomasma hold that in the realm of the lived body, "dis-ease is an interpretation of disruption, an interruption in the ability to cope" (1981, p. 76). Caroline Whitbeck (1981, p. 615) argues that disease (1) is a psychophysiological process, (2) compromises the ability to do what people commonly want and expect to be able to do, (3) is not necessary for doing that which people commonly want and expect to be able to do, *and* (4) is either statistically abnormal in those at risk or people have some other basis for a reasonable hope of finding means to prevent or effectively to treat the process. Charles M. Culver and Bernard Gert state that disease (or as they call the phenomenon, malady) "is a condition of the person that involves suffering or the increased risk of suffering an evil" (1982, p. 71). Lawrie Reznek summarizes the tradition of the foregoing when he says: "Disease is to be understood in terms of the evaluative notion of being harmed. ...[This] leads us to Normativism--the thesis that the concept of disease is value-laden" (1987, p. 170). In short, a prominent view is that disease is not reducible to biological correlates, but rather involves normative judgments concerning ability, function, normality, or harm.

Indeed, matters are not so easily settled. Pointing out that disease is not always something that is *dis*valued or harmful, D. Jennings reminds us that: "One can be seriously diseased without being ill; for example with silent hypertension or an occult malignant disorder" (1986, p. 870). Alternatively, Goosens (1980, p. 102) recognizes that disease can sometimes have benefits and so be valued. Examples such as cowpox providing immunity from the more serious smallpox, sickle cell mutation providing protection against malaria, and an absence of CCR5 genetic trait as a protection against AIDS (National Institutes of Health et al., 1999b, pp. 6-7) illustrate this point. Along these lines, Talbott Parsons (1958) argues that therapy carries benefits such as excuse from responsibility and blameworthiness, where such excuse can bring significant benefits. How values frame disease is a matter of serious discussion and complex matters.

Additional attempts to qualify the character of disease are evident. It has become popular, for instance, to demarcate between disease and illness (Parsons, 1958) in terms of objective and subjective criteria. For Alvin Feinstein, disease is a description of a set of events in morphologic, laboratory terms, while illness is a description in terms of signs and symptoms (1967, pp. 24-25). As Barondess puts it: "Illness is not a biologic, but a human event" (1979, p. 385). One advantage of distinguishing illness from disease is the ability to account for the different meanings of sickness within and between particular cultures. As he says, "Illness is by definition subjective and thus specific to time, place, and culture. Variations in illness are to be expected" (Caplan, 1993, p. 240). From this standpoint, illness accounts for the diversity of disease expression and disease accounts for the similarity of certain clinical problems within and across cultures.

In contrast, F. Kraupl-Taylor argues that the distinction between disease and illness is not so simple. He transcends the split between disease and illness when he contends that "so-called causal treatment of patients, as distinguished from mere symptomatic treatment, means that psychological events and clinical symptoms are affected concomitantly..." (1979, p. 77). Similarly, Per Sündstrom develops the position that so-called icons of disease "are discovered in the *integral* clinical situation, where the 'things' that provide content and form for the icons include not only outer reality but also the inner subjective reality..." (1987, p. 185). This inner subjective reality includes "certain basic value-preferences, such as: life is preferable to death, [and] the integrity of the organism is preferable to pain, hope to despair" (1987, p. 185). The point is that the attempt to distinguish disease and illness may be unsuccessful. Much depends on the context (e.g., clinical care, research, public health) in which disease is understood.

We can conclude, then, that there is disagreement concerning the definition and boundaries of disease. In their study of classificatory habits, Campbell et al. (1979) find that 95% of laypersons and 99% of physicians classified infections (e.g., malaria, syphilis, and measles) as diseases. Conditions caused by physical agents (e.g., drowning, fractured skull, and heat strokes) are classified as diseases by only 9% of laypersons and 44% of physicians. 16% of laypersons and 54% of physicians classify conditions caused by chemical agents (e.g., barbiturate overdose, hangover, carbon monoxide poisoning) as diseases. To add to the analysis, the extent to which so-called asymptomatic, sub-clinical, or pre-clinical genetic conditions are disease engenders further discussion.[8]

In summary, there are debates regarding the character and boundaries of disease. The debates concern whether disease identifies a biological concept, normative state of affairs, or cultural construct. Such debates are not simply academic ones, but ones that affect how humans construct the clinical setting and the clinical classifications that are adopted to account for human disease and health and their treatment warrants. A major goal of this study is to reconcile the debates, and in so doing to provide an alternative way to understand disease.

2. FOUR PHILOSOPHICAL DEBATES

Four distinct but related philosophical debates regarding established definitions of disease may be distinguished and will frame the discussions in this work. The first three tie to the three types of presuppositions of disease, namely, metaphysical, epistemological, and axiological. The fourth is a consequence of the first three and demarcates the context of reality, knowledge, and values in disease.

A first debate concerns the metaphysics[9] of disease, the nature of its being. For this debate, the history of medicine sets the stage for contemporary discussions in the philosophy of medicine. On the one hand, ontological conceptions of disease take disease as an entity in itself. Thomas Sydenham (1624-1689) (1981 [1676]), for instance, argues that nature delivers the structure of disease, which is characterized by recurring, natural, and enduring patterns of signs and symptoms. On the other hand, physiological conceptions of disease rely on some understanding of normal functioning of the body and interpret disease as a deviation from that norm. F.J.V. Broussais (1772-1838) (1981 [1828]), for instance, argues that disease does not constitute a thing. Accounts of disease in terms of recurring, natural, and enduring patterns (such as offered by Sydenham) are derived from metaphysical speculations. Medicine must rid itself of such speculations and conceive disease as a relation between and among actual occurrences of the different organs. This debate contrasts *metaphysical realist* and *metaphysical anti-realist* themes.

A second debate is one about the epistemology[10] of disease, about how we know disease. On the one hand, there are those who argue that reason provides access to knowledge of disease. John Brown (1735-1788) (1803), for instance, holds that it is possible by reasoning alone to ascertain the nature of particular disease mechanisms. On the other hand, there are those who argue that sense experience provides access to knowledge of disease. Pierre-Charles-Alexandre Louis (1787-1872) (1835) and Jules Gavaret (1840), for instance, emphasize that speculative theories and logical deduction cannot provide knowledge of the diseases brought about by environmental factors. Rather, clinicians must rely on positive facts, a legacy of Auguste Comte (1798-1857) (1988 [1830-42]). In short, *rationalist* themes contrast with *empiricist* ones[11] and we are faced with coming to terms with questions regarding how we know disease.

A third debate asks whether values must play a role in disease concepts. This debate focuses on the axiology[12] of disease. On the one hand, some argue that disease concepts are value-neutral and can be specified in terms of typical species functions. For example, being healthy for Christopher Boorse (1975) means being a proper specimen of the species to which one belongs, and being diseased is a state of failing to be a proper specimen. On the other hand, there are those who argue that disease is a normative or evaluative concept best understood in terms of harm or threat to well-being. Disease is a disvalued state of affairs. In this camp are included Pellegrino and Thomasma (1981, 1988), Engelhardt (1981 [1975], 1996), and Ruth Benedict (1934a, 1934b). In short, there are major disputes between *neutralists* and *normativists* about the extent to which values play a role in disease concepts.

Related to this third debate is one concerning the nature of the values that frame disease. On the one hand, some argue that the values central to disease concepts are universal or objective. Edmund D. Pellegrino and David C. Thomasma (1981) and Charles M. Culver and Bernard Gert (1982) contend that objective values, independent of individual opinion, are what secure the possibility of trans-cultural interpretations of disease in medicine. In contrast, others (e.g., Benedict, 1934; Szasz, 1961) argue that the values that frame disease are dependent on or relative to individual or cultural interpretations. Cross-cultural accounts of disease are possible not because of objective or essential values, but because certain physical and psychological patterns of function, action, or behavior are recognized by particular groups as undermining the achievement of human endeavors in a commonly shared environment. *Value objectivist* accounts of disease contrast with *value subjectivist* ones.

A fourth debate explores the extent to which a contextual account of disease is relative.[13] On the one hand, there are those who argue that disease is dependent on socio-historical factors. Ludwik Fleck (1979 [1935]), for instance, forwards the view that disease is a function of specific thought-styles and thought collectives. On the other hand, there are those who worry that any socio-historically conditioned view of disease will inevitably lead to a relativist view. Lawrie Reznek (1987) provides an overview of this concern. In short, there is a dispute between *contextualists* and *relativists* of disease. These and related discussions concerning the metaphysical, epistemological, and axiological character of disease provide this inquiry its general framework for reflection.

3. OCCASION FOR INQUIRY

In this philosophical analysis of disease, one stands at the intersection of medicine and philosophy. On the one hand, an analysis of disease, in conjunction with first-person experiences as patients and advocates of patients, has much to gain from philosophical reflection.[14] Medicine considers questions that tie centrally to what it means to be human--as a knower, performer, and valuer. It inquires into the nature of human somatic and psychological health and disease in order to respond to that which patients bring into the clinic. Along with attorneys and theologians, health care professionals traditionally have not been seen merely as technicians, but as professionals with special obligations or duties to individuals as well as society. The learned professions (e.g., medicine, theology, law), after all, are those that can provide an account of themselves (including their methodologies), assume responsibility for their actions (Callahan, 1988), and place themselves within the general concerns of human culture. It is here that philosophy is able to assist. Philosophy offers medicine ways to think through intellectual and practical issues by providing methodologies for analysis, argument, and critique, ones that draw from logic as well as the history of ideas, thus preventing against conceptual blindness.

On the other hand, philosophy has much to gain from work in medicine, for any worthwhile conceptual analysis of life and death entails considerations of claims made by a discipline that studies life (e.g., biology) and death (e.g., thanatology). [15] Such claims play a critical role in philosophy in providing content to discussions. Reflections in ethics and religious studies on freedom and responsibility take on added dimension when considering new knowledge in genetics. Discussions in philosophy of mind and of psychology are greatly enhanced from work being done in neurobiology, neurophysiology, psychiatry, and behavioral genetics. Social and political philosophy takes on practical import and new perspective when considering the allocation of health care resources or policies protecting patient rights. Reflections in professional ethics find guidance from medicine that, since at least the Ancient times, has entertained questions concerning the proper boundaries of human conduct. In short, medicine has much to offer philosophy.

3.1. Philosophy

In endeavoring to study disease, one is introduced to one of philosophy's central tasks, namely, that of aiding a culture or community in clarifying its views of reality and of itself. *Philosophy* (Gr. *philos*, love + *sophos*, wisdom, or love of wisdom) is an attempt to resolve intellectual questions and quandaries about reality, knowledge, the moral life, and the social order.[16] In the context of medicine, for example, one might ask: "How can I understand what is true or correct in medicine and justify it to others?" "How can I understand what is right, good, or virtuous conduct on the part of health

care professionals and among biomedical scientists and justify that to others?"
Philosophy is neither properly construed as an attempt to decide what people usually
hold about true knowledge or about right, good, or virtuous conduct, nor is it an
attempt to determine what viewpoint would be most credible to most people. It is not
a survey. Rather, it is at its core an endeavor to evaluate reasons and to determine
what reasons can or should be credited by impartial, unprejudiced, and non-culturally
biased reasoners, whose interests are in the consistency and force of rational argument.
Though no such culturally unprejudiced, transcendent, or deified viewpoint can be
fully achieved[17], the goal of its achievement can serve as a guiding, regulative, or
heuristic ideal, suggesting a direction to proceed in attempting to clarify one's ideas,
concepts, or values on a subject. This approach, even when it cannot produce final
answers, can at least enable us to progress by providing some tentative answers, and
by offering some reasons to explain why some resolutions to metaphysical,
epistemological, axiological, and cultural quandaries are better than others in terms
of the extent to which claims are defensible.

3.2. Philosophy of Medicine

This analysis of disease contributes to a growing literature in philosophy of medicine.
By philosophy of medicine, I mean a field of scholarship devoted to the study of the
metaphysical, epistemological (e.g., logical, methodological), axiological, and cultural
issues generated by or related to medicine (Schaffner and Engelhardt, 1998, p. 264).
Involved in this scholarship is particular attention to salient concepts in medicine, such
as disease and health, which shape medical theory and practice (Caplan, 1992, p. 73).
Here I concur with Pellegrino (1976, pp. 13ff) that although philosophy of medicine
may share common grounds with other philosophical disciplines (e.g., philosophy of
science, philosophy of technology), the grouping of issues in philosophy of medicine
proves useful for the development of an enterprise especially devoted to studying sets
of issues and problems particular to life and death. Such sets include, for example,
clinical nosology and nosography, decision-making in diagnosis and treatment, and
ethical issues in biomedicine (e.g., abortion, euthanasia, cloning). Yet, philosophy of
medicine offers practical guidance because it has distinct theoretical perspectives. It
raises questions concerning the presumptions and conclusions of knowledge claims
regarding life and death. For example, it asks how medicine understands its major
concepts (e.g., disease) and what are the implications of such understandings. This is
to agree with Pellegrino and Thomasma (1981, p. 22), who state that many of the
metaphysical, epistemological, and axiological issues raised by medicine (e.g., disease,
illness, health) are susceptible to comprehension only by philosophical analysis, if at
all.

     The understanding of philosophy of medicine offered here contrasts with one
offered years ago by Jerome Shaffer, who asserts that philosophy of medicine can be
resolved into philosophy of science and moral philosophy, so that there is "nothing left
for the Philosophy of Medicine to do" (Shaffer, 1975, p. 218). It also contrasts with
the view advanced by Caplan (1992), which holds that medicine is essentially the
science of biology, and philosophy of medicine is (and should best be pursued as)

philosophy of science. What is left out can be assigned to the realm of moral philosophy. Nor is it to assert that philosophy of medicine involves everything, and that it can handle any and all philosophical quagmires (see Pellegrino and Thomasma, 1981, p. 21). Rather, my view is that philosophy of medicine offers a unique framework for discussions of issues and problems particular to medicine and to our experiences as patients and care-providers that integrate important and special issues concerning who and what we are, how we know, and what and how we value as human beings. These conceptual issues are intimately connected to questions concerning how we can and ought to transform or change ourselves. In other words, philosophy of medicine must address the connection between knowing and acting, theory and practice, diagnosis and treatment, science and technology, objectivity and subjectivity, and science and humanities (also see Khushf, 1997). In this way, I part company with Pellegrino, who holds that philosophy of medicine has primarily to do with the philosophical investigation of the clinical encounter with a human being experiencing health or illness, in a setting which involves intervention (1976, pp. 13-18). Instead, I am in agreement with Julius Moravcsik (1976, p. 337), who claims that philosophy of medicine requires broad reflection and integration of several philosophical disciplines (e.g., philosophy of science, philosophy of technology, philosophy of engineering, biomedical ethics, humanities, and religious studies) in light of current developments in health care education, practice, research, and administration. Philosophy of medicine is at its core interdisciplinary.

What it means to be interdisciplinary, to traverse disciplines, given the institutional constitution and methodological and epistemological cores of academic disciplines, is a key aspect of philosophy of medicine. In order to be interdisciplinary, One suggestion (Ceccarelli, 1995) is that some (but not all) methodological differences among disciplinary approaches may be subordinated when there is a common project and expectation for progress[18]. Evolutionary biologists and molecular geneticists, for example, find shared interests in understanding how Darwin's principles of natural selection operate at the molecular level within large populations (Fuller, 1995). Geneticists and ethicists find shared interest in thinking through the extent to which humans are determined (Lewontin, 1991). Philosophers and medical professionals create common ground in discussions of human psychosomatic existence, often resulting in novel ways to understand human existence, ways that are enriched by thinking in the humanities *and* the sciences (Khushf, 1997). In this way, interdisciplinary studies are marked by an attempt to bridge seemingly disparate ways of knowing (Gardner, 1993) and doing.

How and what one decides can be subordinated in interdisciplinary work is a matter of importance. It seems to me that there must at least be a shared sense of vision, values, and expected outcomes in the process of working together. Participants must agree upon the general purposes and value of the discussion and seek to achieve certain outcomes. As an example, health care providers, philosophers, and theologians may agree to think through the grounds for limiting access to critical care medicine that are shared by Roman Catholic clinicians and provide a consensus statement (Engelhardt and Cherry, 2002, pp. 35ff). In this way, philosophy of medicine provides an exemplar of interdisciplinary studies, the result of which are reflections that cannot possibly occur in any other discipline.

This is not to suggest that philosophy of medicine is new.[19] Those from the humanities and sciences have an extended record of interaction. Physician-philosophers from Hippocrates (5th c. B.C.) and Galen (129-215) through Pellegrino (1976) and Engelhardt (1996) have concerned themselves with human health and disease. Hippocrates is noted for rejecting supernatural "explanations" and emphasizing the role of observation in medical discovery, thus associating him with the Aristotelian school of thought. Aristotle, the son of a physician, believes that medicine could aid in philosophic and moral tasks to a large degree (Owens, 1977). Galen (129-215) holds that it is possible to elaborate and to support theories concerning the fundamentals of the human body. Physician-philosophers Avicenna (980-1037) and Maimonides (1135-1204) preserve Aristotelian naturalism alongside the Scriptural idea of the contingency of the world by arguing that any finite being is contingent in itself but necessary in relation to its causes. Physician-philosopher Sextus Empiricus (third c. A.D.) advocates pyrrhonian skepticism, a kind of mental hygiene or therapy that cures one of dogmatism or rashness in all ideas of thought. John Locke's (1632-1704) (1975 [1690]) philosophic work is shaped to a great extent by his medical orientation through his relationship with English physician Thomas Sydenham (Romanell, 1974, pp. 69-91; Sanchez-Gonzalez, 1990). As Isaac Newton (1642-1727) (1999 [1687]) characterizes his work on dynamics as a contribution to natural philosophy, others of Newton's day, for instance, René Descartes (1596-1650) (1972 [1650]), think that a philosophic approach to such basic sciences as physiology, as well as to clinical or applied medicine, would be highly productive. In a sweeping claim, physician Rudolf Virchow (1821-1902) holds that medicine and philosophy share the goal of providing "general laws of the human race" (1981 [1895], p. 190). Physician-psychiatrist Sigmund Freud (1856-1939) (1966) develops the language of psychoanalysis to explain mental illness in a way that fits his observations in the clinic. Physician-philosophers H. Tristram Engelhardt (1985), Henrik Wulff (1981b), Kenneth Schaffner (1993), Lawrie Reznek (1987), and Robert Aronowitz (1998) engage in extensive analysis of concepts of health and disease. Then there are the calls for a reevaluation of taken-for-granted assumptions in medicine. Physicians Thomas Szasz (1961), Ivan Illich (1976), and M. Scott Peck (1978) reevaluate mental illness in light of socio-cultural forces. Physician Christiane Northrup (1994, 2001) and psychologist Joan Borysenko (1996) call for a revision of traditional medicine's approach to disease in light of woman's experiences. Physician Deepak Chopra (1998) envisions a reintegration of spirituality in health care. Thus, although activity in the philosophy of medicine accelerates, [20] it is really a reemerging field.

3.3. Medicine

Some unique features of the philosophy of medicine involve unique features of medicine (L. *medicina*, medicine, the healing art). Broadly speaking, medicine involves not only what physicians do, but also the intellectual and clinical endeavors of doctors of optometry, of chiropracty, of podiatry, as well as of nurses, physician assistants, health care administrators, pastoral counselors, and allied health professionals.[21] This broad-ranging account contrasts with a more restricted account, such as offered by Donald Seldin (1977, p. 40). According to Seldin, medicine is a discipline that subserves a narrow but vital arena. It cannot bridge happiness, prescribe the good life, or legislate morality. Rather, it can bring to bear an increasingly powerful conceptual and technical framework for the mitigation of the type of human suffering rooted in biomedical derangements.

Despite attempts to constrain its focus, medicine is unavoidably a broad enterprise. Medicine can (and often does) refer to the basic sciences (e.g., theories about the way the eye functions) and applied endeavors tied to diagnosis (of, e.g., retinitis pigmentosa), prognosis, and treatment (e.g., low vision therapy). As a result, one may engender puzzles about theories of function and models of disease processes (e.g., as found in physiology, pathology, and genetics) (Schaffner, 1980, 1981, 1993), about the ways in which health care practitioners engage in their diagnostic, prognostic, and therapeutic activities (e.g., the ways internists make clinical judgments) (Feinstein, 1967; Wulff, 1981a, 1981b; Albert et al., 1988), about how patients assimilate clinical information (e.g., Sorenson, 1974; Blaxter, 1983; Berwick and Weinstein, 1985), and how social factors influence the understanding and treatment of disease (Macgregor, 1960; Shuval, 1981; Sassower, 1993). In this way, medicine is, as Engel (1977) puts it, a biopsychosocial discipline[22].

But medicine is more than a study of the human as object, for humans cannot fully be explained in terms of third person language.[23] Purpose, value, consciousness, reflection, fear, and self-determination complicate the laws of medicine. As this study illustrates, medicine must consider the special complexities of human as individual subject of his or her self-perceived history. In doing so, it must correlate the explanatory modes of the physical sciences with the interpretive modes of the humanities (Pellegrino, 1976, p. 15; also see Wulff et al., 1986, Ch. 9; Habermas, 1972; Gadamer, 1992). It must take into consideration te special complexities of the human person as subject interacting with the human person as object of and in science science.

On this view, medicine is rooted in the history and traditions of human thought, reflection, and experience.[24] Ancient Greek medicine (approx. 500 B.C.-500 A.D.) offers a naturalist, as opposed to a supernaturalist (Admundsen, 1990; Lund, 1936; Laín-Entralgo, 1970; Achterberg, 1991), approach to disease. It is as if all one has to do is to look to nature to deliver the structure of disease and instructions for therapeutics would be forthcoming. One could know fully, a notion that receives support in the Medieval Ages (500-1500 A.D.) by Judeo-Christian scholars. For them, the congeniality between the knower and the known is fortified by a Judeo-Christian God. Maimonides (1135-1204), Sextus Empiricus (b. early third century A.D.), St. Augustine (354-430) and St. Thomas Aquinas (1225-1274) understand reality and the known as created by the same God. Supernatural accounts of disease as punishment for sins, possession by the devil, and the result of witchcraft reflect a view of the universe as created by God, governed by natural law, but faulted by evil influence (Wulff et al., 1986, pp. 81-83).

In an attempt to rid science of faulty speculations and to develop scientific certainty, classical modern medicine separates epistemology from metaphysics, explanation from interpretation, reason from emotion, reason from faith, human reason from Divine reason, body from mind, facts from values, and disease from the sick patient. Fueling these divisions is a skepticism regarding non-rational forms of knowledge and their capacity to grant answers. Although this skepticism appears as far back as Sextus Empiricus, it gains wide acceptance with sixteenth- and seventeenth-century scholars, such as Francis Bacon (1561-1626) (1989 [1620]), Galileo Galilei (1564-1642) (1953 [1632]), and René Descartes (1596-1650) (1972 [1650]). Both accept a mechanistic view of the body as a machine (La Mettrie, 1961 [1748]; Robinson, 1976) and an "experimental" method that investigates the empirical workings of the body as part of the natural order.

Contemporary, or so-called "postmodern,"[25] ways of understanding our world arise from prior assumptions and question the legitimacy of a single narrative or answer to our more fundamental questions forwarded by modern scholars. Accounts (e.g., Fleck, 1979 [1935]; Kuhn, 1970 [1962]; Szasz, 1961) regarding the historical, socially constructed, and culturally-determined character of science and medicine gain prominence. Attention to the rights of minority groups (e.g., Jones, 1981), the mentally ill (e.g., Pence, 2000), prisoners (e.g., President's Commission, 1981, pp. 1145-1146), and women (e.g., Wolf, 1996) reinforce critiques of contemporary medicine. Studies on race and ethnicity (e.g., Roberts, 1996), class (e.g., Rothenberg, 2001), and gender (e.g., Tuana, 1988; Cutter, 1997; Tong, with Anderson and Santos, 2000) in medicine and science emerge. Human reason and its categories come under post-modern critiques, and science and medicine are seen as one way of knowing among others (Sassower, 1993, 1995).

Postmodern critiques have spawned a revival of traditions that encourage first-person reflections (e.g., existentialism[26], hermeneutics[27], phenomenology[28]) and previously disenfranchised minority voices.[29] The hope that a single tradition or philosophy can guide us in matters of epistemological and axiological dilemma increasingly fades (MacIntyre, 1981). As Lyotard (1984 [1979]) warns, the grand narrative has been lost and with it the hope that single answers to our questions will be forthcoming. An implication of the loss of a singular way to interpret reality is an opportunity for previously silenced voices to speak out and share their stories.

Nevertheless, all is not fractured. The technological revolution, undergirded by developments in information technology, link previously isolated voices, communities, and cultures. In that connections can be made that engender discussions, there appears to be some level of shared knowledge and values. We know immediately, for instance, where and when disease outbreaks (e.g., AIDS, severe acute respiratory syndrome [SARS], West Nile Virus) occur. There are crusades for world-wide causes that can be treated effectively (e.g., AIDS). Scholars call for a trans-national or global perspective (e.g., World Health Association, 2003; Po-wah, 2002; Tong, with Anderson and Santos, 2001; Fox 2001) on matters concerning life and death. An emphasis on alternative approaches in medicine (Jonas, 1993; Clouser and Hufford, 1993; "Complementary...," 1997) and the interaction among medicine, spirituality, and the world health situation (Cameron et al., 2000; Engelhardt and

Cherry, 2002; Shea, 2001) gains new momentum. In a sense, a "technoscience globalism" marks the beginning of the Third Millennium and finds expression in medicine. In short, discussions in the philosophy and history of medicine provide an expansive background for study and reflection of disease.

4. LIMITATIONS

This study of disease has its limits. It is not to be taken as an exhaustive treatment of disease. Rather, it selectively treats major debates regarding the status of disease in modern and contemporary medicine.[30] It is not meant to be an historical analysis, although it relies on influential discussions regarding disease in the history of modern Western medicine to highlight salient points regarding the nature of disease.[31] In addition, this study is not in any single way a treatise on disease classification[32], clinical diagnosis[33], clinical decision-making[34], or health[35], although these topics are discussed insofar as they provide insight into particular features of disease. To continue, this study focuses primarily on what is so-called "somatic disease" and does not claim to delve in any detail into the character of mental disease or disorder, which commands a unique history, methodology, and set of problems, some of which are shared. Nevertheless, the lessons learned here about disease have clear applications in a study of psychiatric disease, and references are made as such. Finally, the perspective taken in this essay is more often than not from the standpoint of the clinician, even though I am not one. This is the case because much of the discussions that I rehearse regarding the character of disease is provided by clinicians who have the talent to communicate in the humanities (e.g., King, 1982; Pellegrino, 1976, 1979, 1983; Reznek, 1987; Engelhardt, 1996; and Aronowitz, 1998). For sure, more work is needed from the perspective of patients (see, e.g., Tong, with Anderson and Santos, 2000), particularly given the position advanced in this book. Despite its limits, then, this work sets out to provide a geography of debates regarding the character of disease in order to gain conceptual clarity and practical guidance.

5. PROGRAM FOR INVESTIGATION

To recap, this book is centrally focused on how disease is understood and undertaken in modern and contemporary medicine in the West. It sets forth a contextual account of disease. Disease is theory-laden and evaluative in a way that is context-dependent. An implication of this view is that our understanding and treatment of particular diseases may differ.

This work is divided into eleven chapters. This first chapter serves as an introduction to the analysis. The next chapter, Chapter 2, provides a case study of the development of acquired immunodeficiency syndrome (AIDS), setting the stage for a more detailed analysis of disease. AIDS is selected because our understanding of AIDS has changed in the span of two decades (1980-2000), from viewing it as a syndrome to viewing it as a disease in its own right. Here AIDS is not meant to represent all disease types. Such a position would be misleading given the position advanced in this book, that it will not be possible to map out *the* character of disease. Rather, AIDS is used as an example of how disease is understood and undertaken in contemporary medicine, one that might serve as a model for other inquiries. Along with Chapters 1 and 11, then, Chapter 2 serves as a summary of the project. The remaining chapters set forth the particular features of disease and provide the argument at greater length.

Chapter 3 focuses on the metaphysical question "What is disease?". The analysis draws heavily from the history of medicine in order to make conceptual points about the nature of disease. It considers the debate between ontological and physiological accounts of disease. It illustrates how competing accounts of the nature of disease offer limited yet complementary ways to understand disease. It argues in favor of a *limited realist* view of disease; disease is real yet bound up with the frameworks through which we interpret it.

Chapter 4 addresses the epistemological question "How do we know disease?". It draws again from the history of medicine in order to make conceptual points about knowing disease. It recasts the discussion of what is known in medicine in terms of how we know. It considers the debate between rationalist and empiricist accounts of disease. It argues in favor of a *representative realist* view of disease.

Chapter 5 illustrates a rather underappreciated topic in philosophy of medicine, namely, the relation between knowing and treating in medicine. It sets out to fill this gap. It refers back to Chapters 3 and 4 and extends the analysis of disease to include treating patient complaints. It argues that a pure or unapplied theory of disease is indefensible. Knowing and treating are necessarily linked in accounts of disease. Given this, a *practical epistemological* approach to disease makes most sense.

Chapter 6 considers the axiological character of disease and whether values are involved in disease and, if so, what are their nature. It considers first the debate between neutralists and normativists followed by one between objectivists and subjectivists of values. It argues that one's answer to the first debate about value involvement depends on one's answer to the second debate about the nature of values. It argues in favor of a *limited stipulative* account of the values that frame disease.

Chapter 7 extends the analysis in Chapter 6 by providing a geography of the values that inform disease. It argues that four major kinds of values frame disease concepts. Functional values tell us what ideals of function or activity are proper to an organism. Aesthetic values indicate what ideals of form and grace are worthy of achievement. Instrumental values illustrate how we single out phenomena to be manipulated for purposes of achieving a goal, such as the maximization of benefits and minimization of harm. Ethical values tell us how we assign judgments regarding moral praiseworthiness and blameworthiness.

Chapter 8 offers a synthesis of the foregoing discussions, leading us to conclude that disease is *contextual*. Our understanding of disease depends on the context in which it is expressed. This context is constructed of diverse ontological, epistemological, and axiological commitments. In the end, a "*localized*" account of disease is offered. Disease is located in particular contexts fashioned by historical, socio-political, and cultural influences. With these reflections in hand, the analysis come to a close with Chapters 9, 10, and 11, and considerations of future directions for work on disease. The beginning of the twenty-first century has witnessed an explosion in genetic medicine and concerns about women's health. Taking this as a lead, accounts of genetic disease (Chapter 9) and those of gendered disease (Chapter 10) receive attention. Chapter 11 offers a final summary of the project, with additional thoughts about directions for further work.

6. CLOSING

The analysis of disease offered here provides, on the one hand, an overview of major debates in the philosophy of medicine concerning disease and, on the other hand, a defense of the view that disease is both theoretical and evaluative in a way that is context-dependent. The former is offered to those who wish to become acquainted with a rich discussion of the character of disease in modern Western medicine. An analysis of the former has, in my case, led to the latter conclusions of this project.

CHAPTER ENDNOTES

1. For an introduction to discussions regarding concepts of disease, one might start with the best guide to the literature, Caplan et al. (eds.) (1981), which unfortunately is no longer in print. For helpful overviews of key debates, see Caplan (1993, 1997), "Health and Disease" (Reich, 1995, pp. 1084-1113), Reznek (1987), Wulff et al. (1986, esp. Ch. 6), Aronowitz (1998), Nordenfelt and Lindahl (eds.) (1984, esp. Appendix, pp. 151-173), Engelhardt, with Erde (1980), Engelhardt (1996, Ch. 5), and Sigerist (1943).

2. By commitments, I mean that which is explicitly or implicitly presupposed in one's analysis. One of the jobs of philosophers is to uncover and make explicit such commitments.

3. Colorado Revised Statute 10-3-1104.7 limits the use of genetic testing information by insurers. The statute defines genetic testing as a direct laboratory test of human DNA, RNA, or chromosomes used to identify the presence or alterations in genetic material associated with illness or disease. The statute applies to entities that provide health, group disability, and long-term care insurance and are within the Colorado Insurance Commission's jurisdiction. The covered entities are prohibited from seeking, using, or keeping genetic information for any underwriting or nontherapeutic purposes. Violation of the act is an unfair insurance practice subject to Insurance Commission sanctions. The statute provides a private right of action for individuals injured by wrongful use of genetic information, with both legal and equitable remedies available. Additionally, the prevailing party may recover attorney fees. See Cutter (1998) and Fox (1995). For an overview of other state statutes, see Hudson et al. (1995)

　　　　Statutes prohibiting the use of genetic information by insurers became moot with the passage of the Health Insurance Portability and Accountability Act of 1996 (HIPAA) (Public Law 104-191). Public Law 104-191 makes it illegal to count genetic disposition to disease as a pre-existing condition in group insurance policies. HIPAA limits the ability of group health insurers to deny coverage based on "preexisting" conditions. Under the law, these insurance plans may deny insurance based on a preexisting condition only when medical advice, diagnosis, care, or treatment, was recommended or received within the six-month period before enrollment. Most unexpressed genetic conditions would not meet this insurance exclusion requirement.

　　　　Moreover, HIPAA explicitly protects people seeking presymptomatic genetic testing. It does so by forbidding group plans from denying insurance based on genetic information when the person has not been diagnosed with the genetic condition. This creates an important distinction. Someday it may be possible to undergo genetic testing for depression. Doctors may recommend preventive treatments to people with depression but HIPAA would stop an insurance company from using the recommendation as evidence of a preexisting condition. Thus, HIPAA greatly reduces the risk of genetic discrimination by health insurance companies.

4. In March, 1995, the Equal Employment Opportunity Commission (EEOC) (1995) released new Americans with Disabilities Act defining disability. Disability now applies to individuals who are subject to discrimination in the workplace on the basis of genetic information predisposing them to illness, disease, or other disorders.

5. This is not to suggest that the choices are always rational. As this study shows, the forces contributing to how disease is understood and undertaken are varied and complex.

6. In the United States, nearly one out of every seven dollars is spent on some form of health care. In 1998, this amounted to approximately 13.6 percent of the gross domestic product (GDP), or $4,178 per capita. By way of comparison, health care in Canada, Germany, Belgium, and Austria in 1998 respectively represented 9.5, 10.6, 8.8, and 8.2 percent of each country's GDP, or approximately $2,312, $2,424, $2,081, and $1,968 per capita (Organization for Economic Cooperation, 2000). In the United States, this figure continues to increase, despite the attempt by managed care to contain costs.

7. Reductionism occurs when all laws of the "whole" (or more complex situations) can be deduced from a combination of the laws of the simpler or simplest situation(s) *and* either some composition laws or laws of coexistence (Audi, 1995, p. 492).

8. New knowledge in genetics now enables us to distinguish between pre-existing clinical conditions and existing ones, and to predict the occurrence and severity of genetic disease. Anticipation, for example, allows clinicians to forecast the onset and severity of clinical conditions, giving rise to the terms "sub-clinical events" and "asymptomatic illness." These and other considerations are taken up in Biological Sciences Curriculum Study (1996, 1997).

9. Metaphysics (Gr. *meta ta physika*, after those things relating to external nature, after physics) is the philosophical investigation of the nature, constitution, and structure of reality. In this analysis, the emphasis is on the nature of disease, and the question "What is disease?"

10. Epistemology (Gr. *episteme*, knowledge, + *logos*, discourse) is the study of knowledge and justification. In this analysis, the emphasis is on how we know disease and the relation between the knower and known.

11. Although the distinction between rationalism and empiricism is historically suspect, it provides a useful way to talk about major camps of thought in philosophy and medicine.

12. Axiology (Gr. *axios*, worthy, + *logos*, discourse) is the study of values, or a branch of philosophy dealing with the nature and types of values.

13. The debate is not whether disease is contextual but rather how. This is because the resolution of the previous debates preempts a non-contextual view of disease.

14. Recommended discussions include Pellegrino and Thomasma (1981), Englehardt (1996), McElhinney (1981), and Bickel (1986).

15. See Wartofsky (1975, 1976, 1977, and 1992).

16. Much, of course, has changed. In the Ancient times, philosophy was concerned with wisdom, understood as the highest form of truth. This continued through the Medieval period and complemented the search for God's truth. The Modern period holds that reason will deliver answers to our questions, and postmodernity calls into the question the possibility of truth. Nevertheless, there remains a search for wisdom , albeit a changed notion.

17. For more on human limitations, see Chapters 3, 4, 6, and 8.

18. What constitutes "progress" is anything but simple. See Lakatos (1970), Laudan (1977), and Gutting (1980).

19. There is a rich and complex history of philosophical reflection concerning medicine reaching back to the beginnings of Greek philosophy. These can retrospectively be recognized as part of philosophy of medicine. Nevertheless, it was not until the nineteenth century that the expression "philosophy of medicine" gained currency (Schaffner and Engelhardt, 1998, p. 265). Also see Jonsen (1998) and Stevens (2000).

20. The field of philosophy of medicine experiences new growth at the end of the twentieth century. Consider journals such as *The Journal of Medicine and Philosophy* (inaugurated in 1976), *Theoretical Medicine* (formally called *Metamed*, inaugurated in 1979), *Medicine, Health Care, and Philosophy*, and *Kennedy Institute of Ethics* (inaugurated in 1991). There are book series, such as *Philosophy and Medicine* (inaugurated in 1975) and *Clinical Medical Ethics* (inaugurated in 1991), and databases, such as "Bioethicsline," begun in the 1980s. There are courses with the title of "Philosophy of Medicine" that are offered to students at the undergraduate and graduate level. All of this has transpired since the 1970s.

21. For more on this broad-ranging character of medicine, see Engelhardt (1996, Ch. 5). Also see Engel (1977, pp.129-136), Pellegrino and Thomasma (1981, pp. 58ff), and Veatch (1997, p. 5).

22. The biopsychosocial model of disease emphasizes that how a disease affects any one individual requires consideration of psychological, social, and cultural factors and stresses individual variability of a disease (Engel, 1977, p. 132).

23. First person refers to the person speaking (e.g., I read). Second person refers to the person spoken to (e.g., You read). Third person refers to a person or thing other than the speaker or the one spoken to (e.g., He/she reads).

24. This account is admittedly general and highlights dominant themes for purposes of illustrating major conceptual shifts in medicine that influence what and how we know in the clinical setting.

25. Postmodernism is a complex set of reactions to modern philosophy and its presuppositions about the nature of reality and how it is known. See Foucault (1972 [1969]), Lyotard (1984 (1979), and Ermarth (1992).

26. Existentialism is a philosophical and literary movement that comes into prominence in Europe, particularly in France, immediately after World War II, and focuses on the uniqueness of each human individual as distinguished from abstract human qualities. See Camus, *The Plague* (1991) and Sartre, *Nausea* (1969).

27. Hermeneutics is a type of philosophy that addresses questions of interpretation. See Dilthey (2002 [1910]) and Heidegger (1996 [1927]).

28. Phenomenology is a philosophical tradition developed by Edmund Husserl (1859-1938) (and his followers, emphasizing the description of human experience as directed unto objects, in the sense in which thoughts or wishes have objects, even if unreal ones ("intentional objects").

29. One might note the explosion of courses in ethnic diversity and gender studies in higher education at the turn of the Third Millennium.

30. This is not to suggest that philosophers of science offer little by way of reflection on disease. One could consider in greater detail the account by Hempel (1965, p. 398) of why Jones contracted streptococcal infection. Then there is Achinstein's (1983, pp. 74-102) analysis of Dr. Smith's account of why Bill was overcome by a stomachache. Or consider the syphilis-paresis example offered by Scriven (1959).

31. I am indebted to Harlow Sheidley, Department of History, University of Colorado at Colorado Springs, and Roberto Trevino, Department of History, University of Texas at Arlington, for our discussions of philosophy of history and more specifically the methodology of using historical materials in scholarly analysis (also see Carr, 1969; Toews, 1987; Novick, 1988; Harlan, 1989; Elton, 1991).

32. For discussions on clinical classification, see Wulff et al. , 1986, Ch. VI.

33. For discussions on of clinical diagnosis, see Murphy (1976) and Wulff (1981b).

34. For discussions on clinical decisionmaking, see Feinstein (1967), Schaffner (1980), and Wulff (1981a, 1981b).

35. For sustained analyses of health, see Whitbeck (1981), Nordenfelt (1987, 2001), Mordacci (1995), Lennox (1995), and Lafaille and Fulder (1993).

CHAPTER 2

THE DEVELOPMENT OF DISEASE:
THE CASE OF AIDS

Acquired Immunodeficiency Syndrome (AIDS) provides the basis for our initial reflections on the character of disease. It is a relatively familiar clinical phenomenon, much has been written about it, and our understanding of AIDS has changed dramatically in the span of almost two decades (1980-2000), from viewing AIDS as a syndrome to viewing it as a disease in its own right. This chapter takes advantage of this "discovery" and makes three points. First, the more one advances in the process of accounting for disease, the more one deals with entities that are human constructions of thought expressing certain recognizable observables in nature. As an illustration, the first part of this chapter discusses ways in which contemporary medicine has successfully explained AIDS in terms of a syndrome, an etiological agent, and a model (Cutter, 1988). Second, disease explanations seek knowledge for the sake of action. On this, the second section explores how the epistemic (i.e., knowledge-gathering) and non-epistemic (i.e., action-oriented) concerns of disease interact in how we explain AIDS. As one way to link the first two points, the third section discusses the role negotiation plays in fashioning clinical explanation by examining the ways in which socio-cultural forces shape our understanding of AIDS. In short, the movement through the explanatory levels of disease reflects medicine's effort to understand and to be able to control disease in ways that will facilitate their resolution within particular socio-cultural settings. The following is offered as a way to set the stage for a detailed analysis of disease that takes place in subsequent chapters.

This is not to suggest that AIDS stands for all diseases. There are important biological, evaluative, and sociological differences among the infectious disease AIDS (this chapter), cancer of the cervix (Chapter 7), and the genetic condition called Huntington disease (Chapter 9). In fact, as this study shows, a single concept of disease is not forthcoming. Nevertheless, AIDS provides the opportunity for a detailed analysis of a specific disease, the process of which may be modeled when investigating other disease types.

1. EXPLAINING[1] AIDS

1.1. Syndrome Identifications and Symptomatic Treatments

The year 1981 marks the emergence of a medical consensus that a pattern of observable signs and symptoms forecasting nearly inevitable death is occurring in isolated groups in the United States (Centers for Disease Control, 1981b; Grmek, 1990)[2]. The Centers for Disease Control begins to receive reports from physicians in New York and Los Angeles about patients with an unusual form of pneumonia caused by the protozoan *pneumocystis carinii*. Signs, such as skin lesions, lymphadenopathy, and unusual infections, and symptoms, such as fatigue, recurrent fevers, unintended weight loss, and uncontrollable diarrhea, are brought together in an unorganized pattern, or syndrome. This syndrome is seen to occur primarily in the "four H groups," namely, Haitians, hemophiliacs, heroine users, and homosexual males. The majority of cases are appearing in the last group, one composed of individuals viewed as highly "promiscuous" (Centers for Disease Control, 1981a). The problem is shown to be acquired rather than inherited (Selik et al, 1984). Through the recognition of a collection of signs and symptoms recurring in a specific pattern, acquired immunodeficiency syndrome or AIDS becomes a clinical reality--in this case, in 1982, a clinical classification or rubric commonly recognized and employed to account for phenomena or observable findings (Centers for Disease Control, 1982).[3] As the AIDS researcher Michael Gottlieb put it, "the third case [of AIDS] clinched the realization that what I was seeing was something new. I knew I was witnessing medical history, but I had no comprehension of what this illness would become" (quoted in Wallis, 1985, p. 441). I take Gottlieb to mean that something had emerged that would require the formulation of some framework through which to understand and to interpret it.[4]

    Syndrome identification may be considered the first level in understanding the clinical problem[5] that we call "disease."[6] At the level of syndrome identification, clinical observers focus on a pattern of disease phenomena, a characteristic syndrome. By describing what they observe, clinicians attempt to isolate a set of events and to bring sense to it through a diagnostic category or clinical classification, such as "Gay-Related Infectious Disease" (GRID) (Murphy, 1994, p. 13). The diagnostic category in turn functions to forecast or predict future occurrences by providing a template for comparison and contrast. If clinicians come upon a phenomenon with the signs and symptoms that resemble a previous occurrence, they then can claim that they have encountered an additional case of a particular phenomenon. Such was the experience of Gottlieb and others.

    If, in addition, clinicians notice that certain interventions act effectively to improve conditions, they then come into possession of *symptomatic treatments*. Treatment at the level of syndrome identification seeks relief of the complaints for which patients come to clinicians for help. In the AIDS story, early recommendations on treating a patient's fever and loss of body fluids (Conte, 1983) may be considered a form of clinical maneuvering at the level of syndrome identification. In the AIDS story, treatment at the level of syndrome identification also spells relief for the clinician. The clinician has some intervention, including advice, to offer to the patient, thereby performing an expected function as one who administers at the bedside. In syndrome identification, however, clinicians have no reason to believe that they are addressing some underlying basis of the syndrome.

1.2. Etiological Accounts and Preventive and Curative Treatments

Syndromes provide the basis for the development of etiological frameworks that allow signs and symptoms to be related in causal terms. In the AIDS story, and during 1983 and 1984, several research groups focus on retroviruses for clues to the cause of AIDS (Gallo and Montagnier, 1987). After finding antibodies cross-reactive with HTLV-I in homosexual patients with lymphadenopathy, Luc Montagnier and colleagues at the Pasteur Institute isolate a previously unrecognized virus containing reverse transcriptase that is cytopathic or pathological for cord-blood lymphocytes. They describe this pathogen first as the lymphadenopathy-associated virus, or LAV (Conte, 1983), later as immune deficiency-associated virus, or IDAVs, and finally as lymphadenopathy/AIDS virus, or LAV again (Barre-Sinoussi et al., 1983). The French group subsequently reports that LAV is tropic (i.e., has an affinity) for T-helper cells, grows to substantial titer (i.e., a standard of strength of a volumetric test solution), and leads to cell death (Klatzmann et al., 1984; Montagnier et al., 1984).

Across the Atlantic, a group in the United States, led by Robert Gallo of the National Cancer Institute (NCI) (Gallo et al., 1984), identifies the AIDS pathogen as human T-cell lymphotropic virus, or HTLV-III. The acronym is significant, representing the ongoing research projects of Gallo, who claims to have discovered retroviruses HTLV-I and HTLV-II, which are known to infect helper T lymphocytes, the cells depleted in individuals with AIDS (Popovic et al., 1984). HTLV-III is believed to incorporate itself into the T-cells, ultimately destroying them and leaving the organism without immunological protection.

Discussion does not end here. Those working with Jay Levy of the University of California at San Francisco offer yet another account of the AIDS virus, naming it AIDS-associated retrovirus (ARV). ARV is isolated from AIDS patients in different risk groups, as well as from asymptomatic individuals from AIDS risk groups (Levy et al., 1984). Like HTLV-III and LAV, ARV grows substantially in peripheral blood mononuclear cells and kills CD4 T-cells. The Levy group subsequently isolates ARV from genital secretions of women with antibodies to the virus, data consistent with the observation that men can contract AIDS following contact with a women infected with the virus (Wofsy et al., 1986). Thus, AIDS neither can be a homosexual or "gay" plague[7] (Shilts, 1987, p. 352), nor is transmitted randomly. Transmission of the virus is seen to occur almost exclusively by the exchange of certain infected bodily fluids, which are primarily semen and blood, and possibly also saliva and vaginal secretions (Centers for Disease Control, 1983). This means that unprotected intercourse and the shared use of needles by drug addicts and transfusions are conducive to spreading HIV.

The French-U.S. findings reflect the beginnings of a shift from viewing disease as simply a clinical entity or pattern (*ens morbi*), to interpreting it in terms of a causal entity that can unify the appearances and make the pattern intelligible in terms of casual relations (*causa morbi*) (Engelhardt, 1985). In this move to an etiological account, it becomes important to agree on the terms of the debate. In May 1986, and following lengthy discussions, members of the International Committee on Viral Taxonomy reach consensus regarding what to call the transmissible agent (Varmus, 1989). The result is Human Immunodeficiency Virus, or HIV. The naming of the AIDS virus turns on decisions regarding what accords with proper taxonomic design in biology, what is most acceptable to those working with actual AIDS patients in the clinic, and what is least biased in favoring one interpretation of the causative agent over another (Varmus, 1989; Rather et al., 1985).

An etiological account of AIDS has twin advantages. First, the theoretical significance of the shift from a syndrome to an etiological account lies in the provision of a new account that has enhanced explanatory and predictive powers. What becomes clear is that AIDS patients have a normal or elevated level of antibodies in their blood, but lack a normal number of white blood cells known as "helper T-cells," or T-4 lymphocytes, which play a crucial role in making antibodies effective. Furthermore, white blood cells known as "suppressor T-cells" in CD8 lymphocytes (named after the receptors on the membrane that inhibit the antibody system) are present in increased numbers. The immune system is thus severely crippled. This is more evidence that a virus-like organism is responsible for attacking the immune system. This evidence provides greater detail of analysis and greater ability to describe accurately a precise sequence of regularly associated events before they occur, thus affording the ability of clinicians to predict future occurrences.

Second, the theoretical significance of the shift is justified by its pragmatic force: one can isolate causes that recommend themselves as mutable variables. An etiological account of AIDS provides for the development of *preventive and curative treatments*. Such treatments presuppose the development of tests. In this case, the ELISA (enzyme-linked immunosorbent assay) test looks for the presence of antibodies that the body might have developed to fight the AIDS virus if it is present in the body. The test is good on sensitivity, that is, it rarely fails to detect the presence of the virus. However, it is poor on specificity; it has many false positives, indicating that the virus is present when it is not. As a consequence, a positive reaction [8] to the ELISA test is followed up by either a Western Blot test or IFA to test for the presence of the virus.

Strategies for preventing HIV transmission in the mid-to-late 1980s involve curtailing certain behavioral practices, such as unprotected sexual intercourse and intravenous drug use. Strategies for curing HIV infections involve the administration of specific antiviral therapies (e.g., azidothymidine [AZT] or zidovudine [ZDV]), the so-called magic bullets of AIDS.[9] Timetrexate for the treatment of Pneumocystis carinii pneumonia, especially for those who cannot tolerate standards forms of treatment and ganciclovir for the treatment of cytomegalovirus retinitis are also made available. Later developments include the development of a "cocktail" of multiple drugs or combination therapies, which contains two inhibitors for reverse transcriptase (the enzyme that replicates the HIV genome) and a protease inhibitor that prevents the process of viral assembly (Bartlett and Moore, 1998). Work continues on vaccines, where the effective vaccine needs to activate the cell-mediated immune response, in which cytotoxic T-cells destroy cells that are infected with HIV (Baltimore and Heilman, 1998). Aided by an etiological account of AIDS, clinicians search for the most effective and efficient preventive and curative treatments. In the move from classifying AIDS in terms of symptoms to classifying it in terms of its etiology, a pattern that usefully organizes experience emerges. In this way, in the 1980s, clinical AIDS becomes "HIV disease." [10]

1.3. Clinical Models and Clinical Recipes

In the early 1990s, the recognition sets in that the etiological mechanism of AIDS is more complex than initially thought. There is added pressure to understand the mechanism: 10 million individuals world-wide are reported to have AIDS, with more than a million in the U.S. The cause of AIDS, though perhaps necessary for the manifestation of the disease, is insufficient to account for the clinical problem. Something more than a virus is needed for the occurrence of AIDS. An understanding of the level of virulence of the individual strain, mode of transmission, and immunological weakness in the host, among other factors, are needed. The rate, severity, and onset of HIV disease expression in different people (Fauci, 1993) involve more than simply understanding that HIV causes AIDS. Among other explanations, the mechanism of the functional and quantitative depletion of CD4 T-cells includes direct and indirect mechanisms of depletion (Pantalco et al, 1993). Increasingly, a simple etiological account of AIDS, one that emphasizes a single-causal relation between a virus and a disease outcome, appears insufficient to explain clinical AIDS.

Given the limits of an etiological account of AIDS, clinical investigators prepare for the development of a theoretical framework in which relevant variables can be related within nomological or lawlike structure. *Clinical models* result. Clinical models or multi-factorial analyses of AIDS receive greater support among members of the scientific community in the early- to mid- 1990s because they allow the possibility of various understandings of AIDS, depending on what one is focusing on in the course of the illness. Foci (Fauci, 1993) include (1) the patient's genome; (2) a pre-infection stage; (3) the early period of infection or immune activation--often characterized by a lack of overt clinical symptoms--in which HIV disseminates widely in the body and an abrupt decrease in CD4 T-cells in the peripheral circulation is seen. An immune response to HIV occurs, with a decrease in detectable viremia; (4) the median or long progressive period, in which CD4 T-cell counts continue to decrease and clinically apparent symptoms begin slowly to emerge; (5) the decline of the immune function, in which CD4 T-cells fall to a critical level below which there is a substantial risk of opportunistic infection (e.g., *pneumocystis carinii* pneumonia [PCP]); (6) full blown AIDS, characterized by CD4 T-cell concentration below 200 cells per cubic millimeters of blood (where normal levels are 800 cells per cubic millimeter of blood); and (7) the patient's behavioral or social context. Here one can conceive of AIDS in terms of a genetic (1), immunological (2,3,4,5, 6), pathological (2,3,4,5, 6), infectious (3) or public health (2,7) condition, depending on whether one is a geneticist, immunologist, infectious disease specialist, pathologist, or public health official. In short, clinical models provide a variety of ways to understand disease by singling out particular relations of attention and study.

The possibility of dealing effectively and efficiently with disease gives the shift from disease entities to disease models practical importance. *Clinical recipes* emerge. Disease is no longer appreciated merely as syndromes or as an effect of a single causal entity. It is seen instead as relations among numerous intervening variables. The possibilities for treating AIDS thus broaden. Paralleling the foregoing: (1) We may be able one day genetically to alter humans so that HIV no longer compromises the immunological system. Researchers find that some carry a genetic "defect" that protects against AIDS. Such individuals lack coreceptor CCR5 for HIV (National Institutes of Health et al., 1999b, pp. 6-7). The CCR5 mutation is a deletion of 32 base pairs of DNA that alters an important coreceptor on the surface of macrophages. HIV cannot bind to the altered receptor and therefore does not enter the cell. This protection coreceptor is seen in about 9 percent of the Caucasian population but is nearly absent from the Asian and African populations. Early research on the genetic variant indicates that the mutation may have arisen in Northern Europe about 700 years ago, during the European epidemic, an indication that it may have provided protection against infection by *yersinia pestis*, the bacterium that caused the plague. In addition to this deletion mutation in the CCR5 gene, two other mutations have been characterized that provide resistance to infection or slow the progression to AIDS. Unlike the CCR5 deletions, some of these alleles are found at approximately equal frequencies among different ethnic groups.

Beyond genetically altering patients, further options for treating AIDS include: (2) develop a vaccine; (3) suppress the immune system or revise the lymphoid tissues, (4) modify the immune system, (5) stimulate the immune system, and (6) trea full blown AIDS. Finally, (7) it may be that in certain circumstances that social variables (e.g., hygienic practices, sexual behavior, drug use), which presuppose the least technological advances, are easiest and least expensive to alter (Board of Trustees, 1987). Or it may be, as Fauci (1993) observed, that different combination drug therapies (perhaps anti-viral drugs plus compounds that alter the immune system function in different ways, depending on the stage of the illness) may be found to be effective.[11] In short, disease models give rise to clinical recipes.

Alternatively, clinical recipes give rise to clinical models. As an example, in the late 1980s when the first AIDS drug, AZT, was used to treat patients infected with HIV, clinico-scientists observed that eventually patients failed to respond to the drug. Clinico-scientists found that the virus present early in infection differed from that found late in infection, within the same individual. Eventually they pieced together the following series of events: (1) HIV exhibits the highest mutation rate ever documented. The reasons for this are twofold. First, the enzyme reverse transcriptase, which the virus uses to make a DNA copy of its RNA genome, makes a lot of mistakes (mutations).   Second, HIV does not have an error correction system to repair mutations. (2) Transcription errors by reverse transcriptase produce mutations in the gene for reverse transcriptase. (3) These mutations produce variability in the virus population.   (4) Some variants can better survive in the presence of AZT and come to dominate the virus population. Clinical recipes inform clinical models, thus leading to new avenues of theoretical insights and investigation.

The point here is that medicine conceives of disease in terms of relations[12] that are open to a wide range of interpretations and interventions (Chapters 3, 4, and 5). The advantage of this move becomes evident: it can understand disease in terms of a complex of influences and factors, individual expressions, and the available means of intervention. Medicine can understand and manipulate disease in various ways, thus expanding its explanatory and therapeutic power. Such is the current view that calls us to understand AIDS in terms of a continuum, one involving primary infection (or acute infection), seroconversion (the time when the body develops antibodies to the virus), and immne system decline.   Correlating with this are treatment warrants including early intervention, immunosuppresents, and immune boosters.

Indeed, the move from an etiological account to a clinical model has its challenges. The latter, a multifactorial account of disease, leaves open the possibility of differing and at times competing interpretations and interventions of a particular disease. Since the mid-1980s, the mainstream clinical view has been that HIV causes AIDS, a view explicitly promoted by the National Institute of Allergy and Infectious Diseases (NIAID) (1995). Nevertheless, in widely-publicized critiques of this view, scientist Peter H. Duesberg (1987, 1988, 1989, 1991, 1994) asserts that HIV does *not* singularly cause AIDS. Factors such as promiscuous homosexual activity, repeated venereal infections and antibiotic treatments, the use of recreational drugs (e.g., nitrite inhalants), immunosuppressive medical procedures, and treatment with the drug AZT are also *causally* associated with the occurrence of AIDS. Though HIV may play some role in bringing about AIDS, it is not in any way *the* cause. This line of thinking is echoed by the South African Minister of Health at the 13th International AIDS Conference in Durban, South Africa, when he asserted that "poverty, not HIV, causes AIDS" (National Public Radio, 2000). The thinking here is that factors other than HIV are necessary conditions of AIDS.[13]

Taken together, the views promoted by NIAID, Duesberg, and the South African Minister of Health bring into focus the view that AIDS may be conceived in terms of a theory-structure in which variables associated with the problem (e.g., signs, symptoms, cause[s], immunological state, behavior, social and economic factors) are understood in terms of their causal relatedness. Clinical models offer the possibility of selecting out certain relations amidst numerous others that may be studied and manipulated.

1.4. Various Expressions of Disease

As the AIDS story illustrates, disease involves six distinguishable stages that pair off to make three theory-praxis relations or dyads [14] that can be isolated and distinguished (see Chapters 3, 4, and 5). First, *syndrome identification* and *symptomatic treatments* provide a basis for making repeat observations in the clinic and for recommending appropriate treatments that target symptoms. This stage reminds us that disease identification begins with patients and related parties who bring their concerns to the attention of health care professionals. At this juncture and second, it is common to search for an underlying cause or causes, arrive at an *etiological account*, and attempt to develop *preventive and curative treatments* for the problem. But as is soon realized, disease is more complex. The isolation of a single cause often does not lead to simple and effective treatments. And so third, *clinical models* are called for and *clinical recipes* are developed in order to address health-related problems. This should come as no surprise given that contemporary medical schools implicitly pay homage to a multifactorial approach to understanding disease by structuring medical education in terms of distinct, yet overlapping, departments and disciplines (e.g., biopsychology, biomedical ethics), a legacy that physicians carry with them as they pursue their specialties and in turn join various health care teams and discussions. Yet, this is not to suggest that the integration of disciplines is as it should be (Childs, 1994, 1999). Critics say too often specialists fail to consider the interrelation among variables (e.g., physiological or anatomical dysfunction, family history, lifestyle, diet, environment) contributing to disease. A multifactorial approach to disease challenges us to conceive disease in terms of a relation as opposed to a thing.

2. KNOWING AND TREATING IN CLINICAL MEDICINE

Disease as a clinical explanation reflects in part clinicians' epistemic interests in knowing the character of disease. In endeavoring to know, clinicians attempt to understand a clinical problem from the viewpoint of an unbiased observer, so that their findings may be shared with others. Thus epistemic interests are governed by criteria that are designed to screen out personal value judgments from clinical explanations (McMullin, 1979, 1987). These criteria include empirical adequacy, statistical relevance, coherence, simplicity, and elegance (Feinstein, 1967). These criteria are reflected, for example, in the medical community's pursuit of the etiological basis of AIDS.

This view of the nature of clinical explanation forms the basis of a body of literature in the philosophy of science and of medicine, which is reviewed in subsequent chapters. It presumes clinical problems to be purely epistemic or rational accounts in which explainers agree on (1) how to acquire evidence relevant to the problem, and (2) how to reason with the evidence in order to reach a rationally defensible conclusion that will resolve some question or quandary. This sketch applies to how the Centers for Disease Control (1986) explains AIDS through commonly shared rules of evidence and inference. Such explanations are supposedly apolitical and impersonal. In this noncontextual view, it is as if the various psychological, economic, political, and social differences among explainers do not matter. Participants pose as selfless knowers. In terms of such a view, a final and true explanation should be available to answer the question of why AIDS occurs.

This work challenges that posture. Here, disease is more complex than an apolitical selfless scientific view permits. Clinical explanation of disease seeks not only to establish regularities of occurrences among clinical phenomena and to find enlightening and useful models to account for these regularities. It seeks to alleviate pain and suffering. In other words, clinical explanation of disease involves appeal to non-epistemic (i.e., action-oriented) influences (Chapter 5), such as judgments concerning function, beauty, usefulness, and ethical or moral appropriateness (Chapters 6 and 7). For example, note that AIDS involves concerns about dysfunction (e.g., immunological, respiratory, and dermatological) that are considered worthy of change by patients and their health care professionals. The dysfunction serves as a warrant for clinical intervention. It tells us what ideal of activity (e.g., a functional immune system) is proper to an organism (i.e., the patient) and serves as a goal of clinical intervention.

Yet, biological dysfunction is not the only target for change. Aesthetic, instrumental, and ethical considerations raised by AIDS provide reasons to intervene. Concerns regarding the achievement of beauty expressed through, for example, the elimination of rashes and prevention of wasting[15], especially in countries where resources are plentiful, influence whether and how we target AIDS. Wasting individuals are ugly. In addition, AIDS, or the isolation of HIV, is used to establish the means (e.g., combination drug therapies and public health campaigns) to achieve certain ends (e.g., the alleviation of infections, the boosting of the immune response). Here, instrumental concerns tell us how to get from means to end by appealing to a utilitarian calculus, or the maximization of benefit and the minimization of harm.

In addition, concerns regarding moral appropriateness of a condition brought to the attention of a health care provider influence whether and how we target it. In the case of AIDS, judgments regarding the moral permissibility of homosexual intercourse enter into how we respond to AIDS. Consider how conservative evangelical Christian voices react to AIDS (Treichler, 1999). Drawing from the Bible, they claim that God has declared homosexuality to be a sin (Gen 19:1-29; Rom. 1:24-27; 1 Cor 6:10; 1 Tim 1:10) because "the wages of sin is death" (Rom. 6:23). In their view, an acceptance of this view would lead to the elimination of AIDS. As Nash puts it, "[o]ur only true hope resides in God the Father and Jesus Christ" (Nash, 2000, p. 13). Followers of such a view would likely not support major funding for scientific and clinical research on AIDS. Spiritual counseling is instead the recommended treatment.

For another example of moral considerations at play in disease explanations, consider the debate between Duesberg and NIAAD. Duesberg's dissent carries potential negative impact on HIV-infected individuals and on public health efforts to control the epidemic (Cohen, 1994). HIV-infected individuals may be convinced to continue to engage in risky sexual behavior and needle sharing, to forego HIV testing, or forego anti-HIV treatments that can forestall the onset of the serious infections and malignancies of AIDS (Edelman et al., 1991). Pregnant HIV-infected women may dismiss the option of taking AZT, which can reduce the likelihood of transmission of HIV from mother to infant (Connor et al., 1994). Clearly, from the standpoint of NIAAD, moral considerations such as patient rights (in terms of informed consent) and patient and public welfare (in terms of maximizing benefits and minimizing harms) are at stake.

Non-epistemic considerations enter, then, into the ways in which clinical explanations are fashioned. Clinical explanation is marked by non-epistemic factors not simply because of the ways in which values and culture adventitiously influence knowledge claims but because values and cultural interests are an integral part of medicine as an applied science. Though they may be distinguished, then, epistemic concerns are inextricably tied to non-epistemic ones.

3. NEGOTIATING THE WAYS OF CLINICAL EXPLANATION

In some sense, this analysis of AIDS provides a procedural account of disease, a schema for assessing various claims regarding the ways in which we explain the basic features of disease. These features turn on claims regarding what we think disease is, how we know disease, and how we value disease. These ontological, epistemological, and evaluative commitments interact and influence one another, resulting in various accounts of a particular disease. As a consequence, in the framing of a disease, we face choices. We face negotiating clinical reality.

As a disease model, AIDS can be correlated with infectious, immunological, genetics, psychological, and social variables, depending on whether one is an infectious disease specialist, immunologist, geneticist, psychiatrist, or public health worker. The construal will depend on the appraisal of which disease variables are most amenable to manipulation, by whom, for what goal, and at what cost. For example, an infectious disease specialist would decide that the major factor in AIDS includes HIV, which may be altered by viral agents. A geneticist would be interested in the genetic structure of HIV as well as of the specific host in order to design an appropriate therapeutic response to AIDS. A public health worker would find that the major factors in AIDS include life-style habits, which may be altered using social behavioral techniques (e.g., having easy access to condoms). These and other accounts of AIDS are all the products of negotiation about clinical reality.

Negotiation in medicine takes on a formal and informal character. Formal discussions regarding the nature of disease are found in the literature and in special organized forums. Recall, for example, the work of the International Committee on Viral Taxonomy (Varmus, 1989) and its work on naming the AIDS virus. The Committee, composed of those credited with discovering the virus, considered numerous suggestions for names. For each, they asked whether the proposed name was consistent with scientific facts and current practices in taxonomy and nomenclature, the ability of the name to distinguish the virus from those it does not resemble, and the scientific and clinical usefulness of the name. After choosing "HIV" as the name of the virus that causes AIDS, the Committee welcomed dissenting opinion among its members (particularly those by Max Essex and Robert Gallo) to show the public "the reality of disagreement and the willingness of intellectual combatants to compromise" (Varmus, 1989, p. 8).

Discussions regarding what constitutes disease take on an informal character as well. These are found in the offices and laboratories of clinicians and scientists and, in the case of AIDS, focus on what and how to investigate the disease, whether and how to fund AIDS research, and whether and when to protest the lack of attention to AIDS research and treatment. Either way, clinical reality is construed as the outcome of the choices of various communities of individuals.

The more clinical medicine frames explanations of disease that are open to a range of interpretations, the more it is necessary to have common understandings of the significance and consequences of employing clinical explanations as warrants for action. The work of the International Committee on Viral Taxonomy, along with other committees concerned with developing diagnostic and therapeutic criteria, illustrate medicine's concern to develop common understandings and manipulations of disease for use in the clinical setting. Choices are made concerning which group of studies to rely on and what epistemic and non-epistemic criteria are being satisfied. Considerations include, e.g., empirical adequacy, statistical relevance, coherence, simplicity, as well as functional, aesthetic, instrumental and moral considerations. Medicine would make little sense without common ways to interpret disease.

The selection of a clinical explanation as correct or useful turns on the costs of being wrong. Failure to make acceptable cost-benefit, risk-benefit, or harm-benefit determinations may lead to harm, neglect, and waste. To decide on the costs of various mistakes, one must weigh the costs of treatment for diseases that the patient does not have against the cost of not treating afflicted patients. In the case of AIDS, if one adopts standards for treatment that are too lax, one may unduly increase the economic, social, and personal costs of care for individuals as well as the collective whole. However, if one sets the standards too strictly, one will pay the cost in morbidity, mortality, and spread in the population (Hardy et al., 1986). As a consequence, one must decide upon a prudent balancing of over- versus under- treatment.

Such assessments require comparison of the possible benefits and harms involved in the clinical choices at hand. There is unlikely to be initial agreement with respect to such rank orderings. The results are often clinical explanations with non-epistemic overlays. For example, consider the controversies surrounding the determination of the proper standards of exposure to the AIDS virus in the clinical setting (McCray et al., 1986) and the workplace (Sande, 1986). Such explanations of the ways in which HIV is transmitted from individual to individual turn on varying views of what counts as proper scientific evidence and the rules of inference and what risks we are willing to take in order to save lives in occupational settings. In the early stages of AIDS, these discussions are particularly challenging because of limited data on the nature and model of transmission of HIV.

Increased knowledge does not lessen the burden in decisionmaking. In 1996, Douglas Owens et al. (1996) proposes a policy for the voluntary screening for HIV in hospital populations where the seroprevalence of HIV is greater than 1%. The authors calculate that counseling and testing costs between $30 and $70 per person with an annual cost of $90-210 million. Such a program would provide for earlier educational and therapeutic interventions for high-risk individuals. For their calculations, the authors presume a 15% reduction in risky sexual behavior and needle sharing. They also presume that 50% of patients would decline testing. Given these assumptions, the authors conclude that such a policy would prevent approximately 565 HIV infections yearly. The results of early intervention, treatment, and counseling would be 50,500 life-years saved in patients and 13,500 life years saved in their partners, with a cost of $36,600 per year of life saved. If calculated in terms of quality adjusted years of life saved, the cost would be $55,500 per year saved. These and similar deliberations illustrate the role instrumental concerns play in fashioning disease categories.

In addition to instrumental concerns, moral one influence the construction and use of disease categories.   Take the case of diagnosing women with AIDS. Between 1995 and 1996, the incidence of AIDS decreases by 8% among men, but increases by 2% among women (Munson, 2000, pp. 332-333).   The first natural history study of HIV disease in women begins in 1992, and it is in 1993 (ten years *after* AIDS is first reported in women [Centers for Disease Control, 1983]), that the CDC first publicly recognizes that HIV-related symptoms specific to women exist. At that time, the agency modifies its surveillance definition of AIDS by adding invasive cervical cancer to the list of AIDS-defining conditions, along with pulmonary tuberculosis and recurrent pneumonia (Centers for Disease Control, 1992). Conditions that are manifested most frequently in HIV-infected women, e.g., recurrent vulvovaginal candidiasis, pelvic inflammatory disease, and cervical dysplasia, are part of the list of "symptomatic conditions in an HIV infected adolescent or adult" but are not included among conditions listed as "AIDS-defining" (Centers for Disease Control, 1992). Critics (e.g., Goldsmith, 1992; Faden et al., 1996) have argued that if the CDC criteria were to have taken women more fully into account by including these symptomatic conditions in the "AIDS-defining" category, the number of women with AIDS would have doubled, and women would have been treated more effectively in the early years of AIDS.

But even when women are taken into consideration, the moral challenges do not end. Each year some 7,000 HIV infected women give birth to children in the United States. Some 1,000-2,000 of their infants are infected. Of those infected, 50-70% do not acquire the virus until late gestation or during labor. In addition, there is a risk of transmission through breast milk. Prenatal infection is of even great importance in parts of the world where some 30% of women of reproductive age are infected (Frerichs, 1994). Studies show that among infected women on the AZT protocal, only 8.3% transmitted the virus to their children, as compared to 25.3% of those receiving placebo. Such developments raise concerns. How should medicine determine which pregnant women are at risk of infecting their children? Can a high-risk pregnant woman refuse to be tested for HIV? Should all women of reproductive age receive mandatory HIV screening and counseling? What about the long-term side-effects of AZT therapy on children of infected mothers? Is excluding pregnant women from research on preventing the transmission of HIV ethical?

The discussion thus far implies that in resolving disputes about how best to explain disease, it will be impossible to appeal solely to epistemic criteria in order to acquire a single best answer. The ideal is simply unattainable. In explaining AIDS, one is not simply describing organic dysfunction. One is accounting for a disease in terms of a wide range of factors, which include epistemic and non-epistemic considerations. One is forced to choose among various interpretations, to act on some of them, and eventually to pay the consequences of possible misjudgment. Points of agreement and of disagreement are likely to be multiple and complex. In short, clinical reality, in this case disease, is in some sense a negotiated reality (see Chapter 8). This recognition underscores our choices and indicates our responsibilities as individuals who not only study clinical reality in order to know it (Chapters 3 and 4), but know it in order to manipulate it and manipulate it in order to know it (Chapter 5).

It is no wonder, then, that uncertainty and mystery characterize disease. The uncertainty surrounding disease results in "a certain modesty with respect to the limits of our present conception and practices, and to the inevitability of their replacement by perhaps radically different ones" (Wartofsky, 1976, p. 174). In other words, given that we can never know fully and that our knowledge continually expands and evolves, our knowledge of disease is limited and our task at knowing and manipulating disease does not end. For some, this is an unacceptable position. We so often want answers, especially from physicians and scientists. For others, there is a chance to celebrate the human condition, which is one of change and mystery, that breeds the scientific quest to know more fully and the caution by those in the humanities and theology to proceed carefully.

4. CLOSING

AIDS provides the basis for our initial reflection on the character of disease. At this point, we see that disease is a heterogeneous and complex notion, appealing to a wide range of metaphysical, epistemic, and non-epistemic considerations concerning the character of disease, how we know disease, and how we value it. The forthcoming chapters set out to elucidate key debates in greater detail and offer some conclusions.

CHAPTER ENDNOTES

1. "Explaining" is from the Greek *planus*, which means to make smooth or intelligible (Achinstein, 1983). Disease is an explanatory notion in that accounts of clinical problems set out to make intelligible the pathological findings and processes associated with the complaints or actual mal-experiences of patients (Engelhardt and Spicker, 1975).

2. Some scientists believe HIV spread from monkeys to humans between 1926 and 1946 and first appears in Africa in the 1930s.   Others claims that, in 1959, a man died in the Congo of what we now call AIDS. Others report that gay men in the U.S. and Sweden–and heterosexuals in Tanzania and Haiti–begin showing signs of what we now call AIDS. In *And The Band Played On*, journalist Randy Shilts designates Gaetan Dugas (at the 1980 San Francisco gay pride parade) as "Patient Zero," the man whose erotic penchants and compulsion put him causally at ground zero of the American AIDS epidemic (1987, p. 11).

3. The term "AIDS" first appeared in the *Morbidity and Mortality Report* (*MMWR*) *of* the Centers for Disease Control (CDC) in 1982 to describe "a disease, at least moderately predictive of a defect in cell-mediated immunity, occurring with no known cause for diminished resistance to that disease" (Centers for Disease Control, 1982).   The initial CDC list of AIDS-defining conditions includes   Kaposi's sarcoma (KS), *Pneumocystis carnii* pneumonia (PCP), *Mycobacterium avium* complex (MAC), and other conditions.   It has been updated on several occasions, with significant revisions (e.g., Centers for Disease Control, 1985, 1986, 1987, 1992).

4. The phenomena of dealing with so-called "emerging disease" is not new. Consider here Lyme's disease, Legionnaire's disease, Junta virus, Ebola, and Chronic Disease Syndrome (Cowley, with Hager and Joseph, 1990; National Institutes of Health et al., 1999b). A major cause of the emergence of new diseases is environmental change (e.g., human encroachment into wilderness areas and increased human traffic through previously isolated areas). The re-emergence of some diseases can be explained by the evolution of the infectious agent (e.g., mutations in bacterial genes that confer resistance to antibiotics used to treat disease). The re-emergence of some diseases can be explained by the failure to immunize enough individuals, which results in a greater proportion of individuals susceptible in a population and an increased reservoir of the infectious agent. Increases in the number of individuals with compromised immune systems (due to the stress of famine, war, crowding, or disease) also explain increases in the incidence of emerging and re-emerging infectious diseases.

5. By "clinical problem," I mean that which patients bring to the attention of health care professionals. Clinical problems include disease, illness, deformity, dysfunction, impairment, trauma, and injury. There has been debate in the literature regarding the use of the term "clinical problem." Engelhardt, for example, uses the term in a general way to refer to disease, illness, deformity, and medical abnormalities (1996, p. 189ff).
   It has been suggested that not all foci of attention in the clinical setting are problems (Jennings, 1986; Goosens, 1980), and hence, Engelhardt's term is misguided. What critics of Engelhardt and others fail to appreciate is that the phenomena brought to clinicians's attentions are problems from someone's perspective-- whether this is the patient's, a family member's, an advocate's, or an experienced clinician's. Frankly, most patients are not going to spend the time, energy, and money seeking a clinician if they are not worried about their health. It is important to note that clinical problems do not only include what we refer to as somatic ones involving physical pain, but what we refer to as psychological ones involving mental pain, though the two can never be seen to be distinct and separable.

6. The pattern, which we call a "disease," has reasonable stability when the criteria remain sharp, the elements cohere, and its utility in clarifying experience remains high (King, 1981 [1954], p. 117).

7. The term "plague" comes from the Gr. term *plēgē,* a blow or misfortune, and refers in some contexts to divine punishment. According to Susan Sontag (1990), in popular culture, AIDS becomes known as the "gay plague."

8. A positive ELISA test does not mean that you have AIDS; a negative ELISA test does not mean that you don't have AIDS (Centers for Disease Control and U.S. Public Health Services, 1998).

9. AZT is developed in the late 1980s. The recommended dose is one 100 mg capsule every four hours around the clock.

10. In its 1995 report, the National Institute of Allergy and Infectious Diseases (NIAID) uses the term "HIV disease" (1995, p. 9).

11. In fact, in the early to mid 1990s, such combination drug therapies significantly lead to a decrease in HIV infections. According to a Centers for Disease Control and Prevention Study, 1994-1995 was the first time numbers of AIDS cases began to decrease. During the first quarter of 1994, AIDS death rate was 35.2 per 100 person years. In the first quarter of 1995, the rate dropped to 31.2 per 100 person years when physicians started using the antiviral drugs ZDV (also known as AZT) and 3TC in combination. In 1996, protease-inhibiting drugs came into widespread use and death rates plunged. In the first quarter of 1996, death rates dropped from 29.4 per 100 to 15.4; in the first quarter of 1997, death rates dropped to 8.8 (Munson, 2000, pp. 332-333).

12. By relation , I mean a connection between and among references (e.g., commitments, observations, evidence). In that the connection takes place in time, the connection is dynamic and open to change. We will see how disease fits this model in greater detail in subsequent chapters.

13. One is reminded of studies that establish the intimate connection between socio-economic status and health, where the lower the socio-economic status, the poorer the health. See World Health Association (2003).

14. By "theory-praxis dyad" (Gr. *dyas*, two, consisting of two, a pair of units considered as one), I mean the inextricable relation between theory and practice. A "theory" (Gr. *theoriá*) is a set of hypotheses that posits such entities and properties. Various reductionist, eliminationist, and instrumentalist approaches to theory agree that the full cognitive content of a theory is exhausted by its observational consequences reported by its observations sentences, a claim denied by those who espouse realist accounts of theory. By "praxis" (Gr. *prasso*, doing, acting), I mean a broad interpretation of practice, or that which is associated with production or exchange (Marx, 1961 [1844]), deed or affair (Dewey, 1925), and the entwined phenomena of discourse, communication, and social practices (Horkheimer, 1982). As Chapter 5 argues, there is a necessary connection between theory and practice in medicine because how one understands the world is a function of one's interaction in the world, and how one interacts in the world is a function of one's understanding. As human beings, we know by doing, and do by knowing, a topic that is taken up at greater length in Chapter 5.

15. Wasting is, of course, not just an aesthetic concern. It involves as well those regarding function (e.g., wasting often leads to biological dysfunction) and guides the provision of treatments (i.e., those that reverse wasting). There may also be an ethical component in the concern that we may ask whether there is a moral responsibility to provide to those who are in need.

CHAPTER 3

THE NATURE OF DISEASE

This chapter explores the metaphysics of disease, with emphasis on the ways in which concepts of disease reflect what we know in medicine. It considers traditional discussions in the philosophy and history of medicine about the nature of disease, with particular attention to its structure and development. It argues that competing accounts of the nature of disease offer limited yet complementary ways to understand the metaphysical character of disease, thereby resulting in what is called a *limited realist* approach. As Khushf (1995, p. 465) rightly notes, coming to terms with the tensions between and among diverse ways of conceptualizing disease is perhaps one of medicine's important goals as it moves into the Third Millennium, a time that promises great strides in our understanding and control of disease [1].

1. WHAT IS DISEASE?: A DEBATE

The debate over whether disease entities are natural kinds or artificial clusters of findings fashioned on the basis of human purpose has been characterized in the history of medicine as a debate between "ontological" and "physiological" approaches to disease.[2] In these debates, distinctions are drawn at times between anatomical and physiological accounts of disease, at times between realist and nominalist accounts of disease, and at times between rationalist and empiricist accounts of disease. This chapter focuses primarily on the first two debates; the third is taken up in the Chapter 4. The dispute in the first debate roughly is whether disease is in some sense a substantial thing or primarily relational in character. Ontological or essentialist theories frame views within which diseases are appreciated as enduring non-conventional patterns. Physiological or nominalist theories frame views within which diseases can be appreciated as departures from general regularities, which are selected in terms of instrumental concerns. In the first case, the accent falls upon the disease as thing; in the second case, the accent falls upon the individual and his or her circumstances, including the laws of physiology, and what one wishes to hold as medically relevant sufferings. In this way, the notion of disease as a thing or state contrasts with that of disease as a relation or process. Consideration of the debates between ontological and physiological accounts of disease leads to an appreciation of what is disease.

Some might say that there is no need to consider the metaphysics of death because few talk about it anymore. Others might state that with the rise of the modern project, metaphysics ceases to have import in our discussions. This chapter challenges these claims by laying out the historical debate and illustrating the relevance of the

32

discussion today. Metaphysics is not dead; it is just underappreciated. Any question concerning "how we know?," which is the topic of the subsequent chapter, assumes an object of knowing. It is the nature of this object that is the concern of metaphysics. In the case of disease, how we know disease is a function of what we think is disease. In addition, history is not irrelevant. It illustrates the paradigms [3] we have previously adopted in order to ground our thinking. These paradigms do not die but are changed in order to give way to new ones, which often have features of the former. It is important, then, to consider the ways in which our understanding of disease has evolved.

1.1. Disease as Substance

Ontological approaches to disease understand disease as substance, as a thing or state. There are at least two major accounts of this view.[4] First, disease is understood as an unchanging pattern of signs and symptoms. Second, disease is understood as a physical entity, which has two major expressions, disease as an internal agent and disease as an external agent. The analysis begins with the first account.

1.1.1. Syndrome as Disease

Our review of the development of HIV disease or AIDS in Chapter 2 begins with understanding AIDS in terms of a *syndrome*, a constellation of signs and symptoms. Signs, such as fever and weight loss, are brought together with symptoms, such as aches and lethargy, and organized into an initial clinical classification in which a comparison and contrast between physiological dysfunction is possible. In defining disease as syndrome, there is a temptation to see disease as a static, unchanging set of clinical events, as a real entity. This temptation has historical roots, as seen in the works of English physician Thomas Sydenham (1624-1689).

Sydenham is widely recognized for first developing an ontological account of disease in early modern medicine.[5] In *Observationes medicae* (1981 [1676]), Sydenham sets forth a method for describing and classifying disease in terms of symptom constellations. Sydenham separates patient symptoms into two categories: (1) pathognomonic symptoms--those that are shared among groups of patients, and (2) idiosyncratic symptoms--those whose pattern and character are unique to a particular patient. Sydenham groups the pathognomonic symptoms into named disease entities, which he believes are as real and tangible as other creations of nature. Sydenham's project presumes that the symptoms of patients coalesce in natural, enduring patterns, or syndromes. As he says, "...diseases can be reduced to definite and certain *species*, and that with the same care which we see exhibited by botanists in their phytologies...." (Sydenham, 1981 [1676], p. 146)[6]. The description should be a systematic ahistorical account of the phenomena of disease.

> In writing the history of disease, every philosophical hypothesis whatsoever, that has previously occupied the mind of theauthor, should liein abeyance. This being done, the clear and natural phenomena of the disease be noted--these,and these only....No man can state the

errors that have been occasioned by these physiological hypotheses (Sydenham, 1981 [1676], p. 147).

According to Sydenham, problems and complaints of patients are to be described phenomenologically as they are presented with as little, if any, influences by previous presuppositions. The typologies of disease, i.e., the nosologies, are thus not inventions, but discoveries in experience, found in the phenomena themselves. As Sydenham puts it, one finds the patterns presented in nature:

> ...Nature, in the production of disease, is uniform and consistent; so much so, that for the same disease in different persons the symptoms are for the most part the same; the selfsame phenomena that you would observe in the sickness of a Socrates you would observe in the sickness of a simpleton. Just so the universal characters of a plant are extended to every individual of the species; and whoever (I speak in the way of illustration) should accurately describe the colour, the taste, the smell, the figure, &c., of one single violet, would find that his description held good, there or thereabouts, for all the violets of that particular species upon the face of the earth (Sydenham, 1981 [1676], p. 15).

One can, then, according to Sydenham, discover real and recurring disease patterns in nature.

Despite suggestions to the contrary (Sydenham, 1981 [1676], p. 132)[7], Sydenham remains committed throughout his life to classifying clinical problems without reliance on abstract etiological factors or background presumptions. Sydenham stands in the scientific tradition of Baconian experimental philosophy in which "[o]ne method of delivery alone remains to us...: we must lead men to the particulars themselves [i.e., facts], and their series and order..." (Bacon, 1989 [1620], p. 48). Baconian empiricism presumes that (1) the scientist has available a well-defined method, a series of steps that if followed necessarily lead to truth, and (2) science is not a matter of probability or approximation of theological or political persuasions but of facts that can be known with certainty. If the method is followed, certain and necessary truths (i.e., facts and causes in Nature) can be attained (Bacon, 1989 [1620], pp. 47-52). Accordingly, Sydenham maintains that medical knowledge can be built on the observed manifestations of disease. Yet, contra Bacon, he is skeptical about the individual's ability to know causes in Nature (Syndenham, 1981 [1676], p. 151). For Sydenham, causes are not observable in nature. True knowledge of disease is founded on recorded observations of signs and symptoms, which are supposedly theory-neutral.

Under the influence of Sydenham's method for providing natural histories of disease, eighteenth-century clinicians such as Carolus Linnaeus (1707-1778) (*Genera morborum* [1763]) and Francois Bossier de Sauvages de la Croix (1707-1767) (*Nosologia methodica* [1768]) classify complaints undistorted by various theories of disease in order to give a basis for prognosis. In *Genera morborum*, Linnaeus claims that diseases can be distinguished on the basis of three criteria: causes, effects, and signs (1763, p. 3). Since one is unable to know with certainty the interactions taking place in the body (i.e., the causes and effects of processes), diseases are to be classified on the basis of symptoms or external signs. The result is a descriptive classification or nosology consisting of eleven classes (*Figure 1*):

Class I.   *Exanthematici* (exanthema)
e.g., smallpox, measles, syphilis

Class II.   *Critici* (critical fevers)
e.g., continued and intermittent fevers

Class III.   *Phlogistici* (inflammations)
e.g., pleurisy, hepatitis, nephritis

Class IV.   *Dolorosi* (painful diseases)
e.g., abdominal pain, arthritis

Class V.   *Mentales* (mental disturbances)
e.g., delirium, rabies

Class VI.   *Quietales* (impaired or lost voluntary actions)
e.g., fainting, paralysis, anorexia

Class VII.   *Motorii* (convulsive diseases)
e.g., hysteria, epilepsy

Class VIII. *Suppressorii* (suppression of bodily fluids)
e.g., asthma, constipation, ammenorhea

Class IX.   *Evacuatorii* (discharge of fluid)
e.g., diarrhea, diabetes, gonorrhea

Class X.   *Deformes* (physical wasting)
e.g., rickets, scurvy

Class XI.   *Vitia* (skin diseases)
e.g., tumors, fractures, wounds

Figure 1. Linnaeus's Eleven Classes of Disease (*Genera morborum* [1763], translated by Bowman, 1976a, p. 9)

Conditions such as fever (Class II), pain (Class IV), and diarrhea (Class IX) are regarded as diseases in their own right. Following closely the procedure of botanists (King, 1982, p. 110), Linnaeus hopes to arrive at a clear knowledge of disease, and thus raise medicine to the same clarity as that of botany, absent intrusions of underlying pathological presuppositions and speculative distractions.

Class I.   *Vitia* (injuries of skin)
              diseases of little importance to the physician,
              to be treated by the surgeon, e.g., sarcoma (tumors),
              varix (dilated veins), punctures.

Class II.   *Febres* (fevers)
              diseases accompanied by fever, chills, and rapid
              pulse, e.g., continued and intermittent fevers.

Class III.   *Phlegmasiae* (inflammations)
              diseases accompanied by fever, inflammation and
              skin eruptions, e.g., smallpox, hepatitis, measles

Class IV.   *Spasmi* (convulsive diseases)
              e.g., tetanus, epilepsy, hysteria

Class V.   *Anhelationes* (respiratory disturbances)
              e.g., asthma, dyspnea (difficult respiration)

Class VI.   *Debilitates* (weakness)
              diseases accompanied by impaired or lost strength,
              affecting cognitive, motor and sensitive functions,
              e.g., syncope (fainting), apoplexy, anorexia (loss of
              appetite)

Class VII.   *Dolores* (painful diseases)
              e.g., heartburn, abdominal pain, wandering pains

Class VIII.   *Vesania* (mental disturbances)
              diseases characterized by derangement of judgment, e.g.,
              delirium, hypochondriasis, nostalgia.

Class IX.   *Fluxes* (discharge of bodily fluids)
              e.g., hemorrhage, diabetes, diarrhea

Class X.   *Cachexiae* (physical wasting and deprivation)
              diseases characterized by changes in
              appearance, e.g., rickets, leprosy, chlorosis
              (discoloring), pregnancy

Figure 2. Sauvages's Ten Classes of Disease (from *Nosologia methodica* [1768, pp. 92-95],
translated by Bowman, 1976a, p. 8)

Similarly, Sauvages details an extensive range of patient complaints through a ten-fold classification of disease (*Figure 2*). In *Nosologia methodica* (1768), Sauvages attempts to erect a typology of diseases on the basis of Sydenham's presupposition that clinical findings fall into easily identifiable constellations of signs and symptoms that can be recognized through their natural histories (King, 1966). His nosology follows:

What Sauvages undertakes is primarily a description of the world of illness, of iatrotropic states, i.e., those states of affairs that individuals bring to the attention of physicians. Conditions such as fever (Class II), pain (Class VII), and fluxes (Class IX) are regarded as disease in their own right. The result is in 1768 the division of ten classes into forty-two orders, three hundred fifteen genera, and two-thousand-four-hundred species, depending on which edition one consults (Bowman, 1975a, p. 5). Under the influence of Sydenham's concept of natural histories of disease, medicine understands disease phenomenologically, in terms of syndromes, observable constellations of signs and symptoms.[8]

Sydenham and the early modern clinicians' influence on how we understand disease is evident today. As an example, popular clinical reference books produced for parents (e.g., Schmitt, 1999) use signs and symptoms as ways to understand disease. Schmitt, for instance, organizes "each of the topics or diseases" beginning with "symptoms and characteristics" (1999 p. xvii). When faced with the task of identifying a disease, he asks his reader to identify "What are the symptoms?" or "What does it look like?" (1999, p. xvii). As an example, one might wonder what are the spots on a child's hand? A closer inspection indicates that the spots are red, pink, and slightly bumpy. A symptomatic report indicates that they are not itchy but the child is "hot." Such signs and symptoms lead to the diagnosis of "hand, foot, and mouth" disease (which is caused by coxackie A virus). Such an approach is useful for organizing clinical phenomena and continues to play an important role in the home and clinic. Patient reports and observation remain key in piecing together a clinical picture.

1.1.2. Physical Entity as Disease

Ontological approaches to disease are adopted as well by those who subscribe to the view that disease is a real internal or external substance, or an etiological agent. One notes how this second group of ontologists differ from the first. For the first, disease is a syndrome, a pattern of signs and symptoms. For the second, disease is no longer identified with a cluster of signs and symptoms. Instead, disease is identified with its cause or etiology--and the cause is located either in the internal or the external environment.

Disease as Internal Substance

As Chapter 2 illustrates, one way to understand AIDS is in terms of internal substance. A focus on the genetic protection against AIDS highlights this orientation. Those with a CCR5 mutation on the surface of macrophages are found to carry a genetic defense against AIDS (National Institutes of Health et al., 1999b, pp. 6-7). Alternatively those without the mutation are vulnerable to AIDS, leading to a drop in an individual's CD4 T-cell concentration below 200 cells per cubic millimeter of blood, where normal CD4 T-cell levels are 800 cells per cubic millimeter of blood (Centers for Disease Control and U.S. Public Health Service, 1998). Here the focus is on the internal substance of AIDS (a genetic mutation), one that has roots in the history of modern medicine.

Friedrich Hoffmann (1660-1692), for example, holds that disease is a state in which internal vital actions (i.e., powers, forces) are distorted and impaired (1971 [1695], II, 1, 1). Disease, as the opposite of health, consists in a great change and disturbance in the proportion and order of the motions, along with which there is a "striking disturbance of secretions, excretions, and other functions of the living body" (quoted in King, 1982, p. 134). As Hoffmann remarks, disease causes must be sought not in the excess or deficiency of certain humors, but rather in the abnormal composition and motion of the blood. "Very many diseases seem to arise," he observes, "when the circulation of the animal spirits and blood, through minute vessels, is impaired and uneven" (1971 [1695], II, 1. 28). Reminiscent of Descartes (1972 [1650]), Hoffmann regards the body as a machine that follows the laws of mechanics, which governs particles in motion. Life results from causes that are wholly mechanical, or reducible to the motion of particles.

Likewise, Hermann Boerhaave (1668-1738) defines disease as a state in which internal vital functions are impaired or disrupted (1742-1746 [1708], sec. 696). "To know disease," he said, "is to know the defect of the Function. The cause of disease is often to be sought in the bodily fluids, more specifically, in the varying quantity and quality of particles in the blood" (1742-1746 [1708], sec. 775). While the particles of blood in a healthy body with normal circulation are spherical, they assume other shapes (e.g., "acute solid angles" [1742-1746 (1708), sec. 725]) in less favorable conditions. More specifically, an increase in the size or density of the particles gives rise to dilation, rupture, and obstruction of vessels (1742-1746 [1708], sec. 721-7 23). This tendency to incorporate the principles of the physical sciences into the life sciences further reinforces the acceptance of Cartesian mechanical philosophy.

Hoffmann and Boerhaave represent an early tradition of thought that holds that disease is reducible to physical substance in the body (Lindeboom, 1968). This thinking forecasts nineteenth century thinking in pathology as well as twentieth century work in genetics. Cellular pathologist Rudolf Virchow (1821-1902), for instance, holds in his later writings that "disease is not in the blood, rather the disease is the effect of the cause on the cells (tissues)" (Virchow, 1981 [1895], p. 195). Disease is for the later Virchow basic ingredient changes at the cellular level. "Most diseases are not elementary processes but rather compound processes where alterations of several or many cell territories co-exist or range themselves alongside each other" (Virchow, 1981 [1895], p. 191). In this way, Virchow labels his own view "clearly ontological" (Virchow, 1981 [1895], p. 191).

Current work in genetic medicine fosters the view that disease is internal substance. With increasing control of environmental factors such as infection and malnutrition, we come to see genes as playing a more significant role in disease (e.g., stroke, coronary artery disease, schizophrenia) in the developed world. As Weatherall says, "It is now clear that many of the major diseases of unknown cause that affect western societies...have an important genetic component, and that many forms of cancer are due to inherited or acquired changes in the genetic make-up of cells" (1991, p. 1). New knowledge in genetics (Chapter 9) focuses our attention on what we inherit from the germ cells of our parents and the acquired changes in cells of any organ during an individual's lifespan. On this view, disease is internal substance, in this case, a gene or genes.

Disease as External Substance

In contrast, there are those who hold that disease is that which invades the human body, a foreign substance. Disease is identified with an external agent or cause, such as a virus, bacteria, parasite, or fungus. An illustration is the identification of AIDS in terms of HIV (National Institute of Allergy and Infectious Diseases, 1995, p. 9). The view is that HIV establishes itself inside a T-cell lymphocyte, a major component of the body's immune system. The virus uses the host cell's reproductive system to replicate itself many times. In this way, HIV eventually destroys the host cell and moves on to infect other T-cells. The body is thus left vulnerable to opportunistic infections, which are easily fought off by a normally functioning immune system. The infections become highly disabling and potentially fatal if the infection is not controlled (Centers for Disease Control and U.S. Public Service, 1998).

The notion of disease as external substance is not new. In 1530, Girolamo Fracastoro (1484-1553) suggests in a poem that syphilis and other diseases could be contagious--that is, they could be transmitted by contact with an infected person, contaminated materials, or infected air (Garrison, 1929, pp. 232-233). The discovery of microorganisms by Anton van Leeuwenhoek (1632-1723) lead some to speculate that the microscopic organisms might be the key to disease. Although this "germ theory" of disease is first proposed in 1762, it is developed by Ignaz Phillip Semmelweis (1818-1865) as he investigated the spread of childbed fever in the obstetric wards in Vienna. His careful observations lead him to conclude that the agents responsible for childbed fever (leading to maternal death) are cadaveric particles, that is, morbific or dead materials, passed from the sick and dead to healthy women on the hand of medical attendants (Hudson, 1983, pp. 146-153). An etiological account of disease is fully developed by Robert Koch (1843-1910) (1932 [1882]) as he studied anthrax, a disease of cattle and sometimes humans. Following Koch's work, scientists identify the bacterial cause of many human diseases (e.g., cholera, glanders, diphtheria, typhoid, gonorrhea, pneumococcal pneumonia, Malta fever, meningococcal meningitis, tetanus, plague, botulism, acute dysentery, and numerous other infections due to staphylococci and streptococci) (Hudson, 1983, Ch. 8). Such discoveries lend great support for the view that disease is an external substance.

1.1.3. Metaphysical Realism and Natural Kind Accounts

"Ontological" construals of disease presume for the most part metaphysical realist presumptions. To talk about illness as reducible to substance (e.g., syndrome, internal cause, external cause) is to talk in a metaphysical realist mode with an emphasis on universals with abstract essences. Metaphysical realism is the view that objects of knowledge have external existence, independent of our experience. More specifically, objects have properties and enter into relations independently of the concepts with which we understand them or of the language with which we describe them. In the case of disease, understanding disease, as portraying natural or logical disease types, best illustrated by classical cases and obscured by atypical ones, provides a backdrop for contemporary interpretations of disease as a natural kind. [9]

Consider, for example, a recent metaphysical realist defense of the view that there is something substantial that we can say about disease. Robert D'Amico argues that, first, a "flexible philosophical account of natural kinds, as found in Mill for example," (1995, p. 566) can ground a concept of disease. This account "properly separates what is a matter of how the world is from what humans value and then variously choose to prefer" (1995, p. 566).[10] For Mill, "kinds in nature" is defined as those classes of objects that "differ from one another not by a limited and definite, but by an indefinite and unknown, number of distinctions" (1874, p. 408). A kind is a set of properties held together by a "uniformity of co-existence between properties" (1874, p. 408). Disease is a natural kind first because the emergence of a medical expertise for the identification of disease lends support to the naturalistic account. Second, disease is a natural kind because "[t]he diverse systems of classification in the history of medicine proceed not, in most cases, from mere convenience or arbitrary choice, but from a first reliance on descriptive accounts worked out prior to some access to the underlying causal processes of these conditions" (D'Amico, 1995, p. 557). Once the underlying nature (e.g., a specific virus or bacteria) is isolated, the previous descriptive account is either reformed or adjusted to the now discovered underlying disease process. In short, for D'Amico, disease names a natural kind: diseases share the same underlying nature, which is an object of scientific inquiry.

1.2. Disease as Departure from Physiological Norms

In contrast to ontological approaches, physiological approaches to disease underscore the individuality of disease states because every disease state can be understood in terms of departures from general physiological or functionalist norms. Physiological views are anti-ontological. Disease is neither an observed set of signs and symptoms nor a perjuring type of pathology. It is not a thing. Rather, for nosologists, i.e., classifiers of disease, disease is relational rather than substantial, the result of individual constitutions, the laws of pathology, and the peculiarities of environments. As a consequence, there are, at least for some, no discoverable absolute borders between disease and health, abnormality and normality, and the pathological and the healthy (Canguilhem, 1978 [1966]).[11] Disease is quantitative and qualitative variation rather than a thing. This view accommodates the shifting, indefinable point at which health becomes disease, and disease health, as well as the broad margins at which there is disagreement as to what constitutes disease (e.g., drug addiction [Leshner, 19970], attention deficit disorder hyperactivity [ADDH] [Kauffman, 2001], premenstrual syndrome [PMS] [Northrup, 1994]).

1.2.1. Conceptualizing Relations: Disease as Process

A physiological conception of disease provides a variety of ways to understand AIDS by singling out particular relations of attention and study. Whether one conceives AIDS in terms of a genetic, immunological, pathological, infectious, behavioral, or public health condition depends on how the relations among the variables that make up disease are understood. To understand a clinical condition as a genetic disease is to focus on the relation among genes, proteins, and the environment. To construe a clinical condition as an infection is to focus on the relation between an external agent and its result. To understand a clinical condition as a public health issue is to focus on the relation between a clinical event and its public health consequences. Much turns on what clinical relations are singled out, why, and for what purposes.

One of the first most influential formulations of the pathological view of diseases in the modern era appears in the 1761 work of the Italian anatomist Giovanni Battista Morgagni (1682-1771), *De sedibus et causis morborum per anatomen indagatis libri quinque* (*The Seats and Causes of Diseases Investigated by Anatomy*) (1981 [1761]). As the title indicates, Morgagni understands each disease to have an organized seat or anatomical resting place in the organs of the body. The structural changes are unique, like fingerprints, and allow those who inspect them to determine the patient's disease. Morgagni's work understands disease as structural changes that are reminiscent of the views of Theophile Bonet (1620-1689) (cited by Morgagni, 1981 [1761], p. 157). As Morgagni says:

> For finally, in respect to my own observations, I have particularly related in each, the year, month, and place in which they were made, and who assisted me, or were present, at the time, unless I had sufficiently done it before. And I have not only remark'd the age and sex of the patient, but other things..., as far as it was in my power to learn, and amongst these such as relate to the method of cure which had been applied... (1981 [1761], p. 164).

While Bonet indeed examines the marks of disease in the viscera (i.e., organs) of dead bodies, it is by the vast scope of his work and its careful correlations of clinical symptoms and autopsy findings that Morgagni makes pathology a genuine branch of modern science and medicine and introduces the "patho-anatomical" idea into medical practice (Garrison, 1929, p. 354).[12]

Following Morgagni, Theophile-Hyacinthe Laennec (1781-1826) in *Traite de l'auscultation mediate* (*Treatise on Medical Auscultation and the Diseases of the Lungs and Health*) (1834 [1819]) outlines a methodology that is to guide the pathological anatomist in understanding disease. The proposed methodology involves (1) identifying a pathological condition in the cadaver through physical changes in the organ, and (2) recognizing the same pathological condition in the living, if possible, through physical signs independent of symptoms, which are unreliable (Faber, 1923, p. 35). In observing the gradual pathological changes taking place in the course of the disease, Laennec recognizes for the first time that phthisis (which is later called pulmonary tuberculosis) is a disease different from other diseases of the lung (e.g., abscess, anthracosis, cancer) (King, 1982, pp. 34-35; Akerknecht, 1982, pp. 93-94)

Reminiscent of Descartes and seventeenth-century mechanists, Xavier Bichat argued in *Anatomie general applique a la physiologique et al medicine* (*General Anatomy Applied to Physiology and Medicine*) (1801) that all animals are composed of organs that are like parts in the machinery of the body. The organs are formed by particular tissues in the same way chemical compounds consist of chemical elements. The tissues are divided into nineteen observable varieties: cellular, nervous, arterial, venous, exhalant, absorbent, osseous, medullary, cartilaginous, fibrous, fibro-cartilaginous, muscular, mucous, serous, synovial, glandular, dermoid, epidermoid, and pileous (Bowman, 1976c, p. 4). These are indivisible parts, like the elements in chemistry, each having its own particular kind of sensibility and contractibility. Bichat (1771-1802) interprets the alterations of tissues or membranes as the meaning of disease, an idea that he attributes to Phillippe Pinel (1745-1826) (1798). In this way, Bichat anticipates the thinking of Jacob Henle (1809-1885) (Garrison, 1929, p. 457) and the early histologists.

Bichat develops a universe of pathoanatomical observations that he hopes, by relating these to the world of the clinician, would effect a reclassification of the world of clinical description and provide a better account of its structure. In *Pathological Anatomy*, Bichat remarks:

> In local diseases the manner of proceeding must be different: we must first examine the affected organ, then the neighboring organs which participate in the lesion; and afterwards proceed to examine the functions. The advantage resulting from this method is, that we may narrowly sift the maladies which affect every system. By this means, the diseases of two organs utterly different will not be confounded, although situated in the same cavity. However, there are some affections which do not admit this methodical classification: such as scurvy, syphilis, &c.; but a persevering study may enable us hereafter to find for them a fit place (1981 [1827], p. 170).

For Bichat, disease results from morbid changes taking place in the various and specific tissues that make up the particular organ. Each tissue has different vital[13] properties and exhibits different sets of symptoms. As Bichat himself acknowledges, challenges to this theory include diseases that are not localized (e.g., scurvy, syphilis). Nevertheless, there is the hope that anomalies would someday be understood in light of this approach.

F.J.V. Broussais (1772-1838) (1981 [1828]) follows Bichat (1801) in initiating a conceptual shift in pathological thinking. In *Anatomie general applique a la physiologie et la medicine* (*General Anatomy Applied to Physiology and Medicine*) (1821), Broussais uses the term "ontology" to refer to disease classifications such as Pinel's (1798), that depict diseases as essences connecting clinical symptoms. Such classifications are to be rejected because they

> have filled the nosological framework with groupings of arbitrarily formed symptoms,...which in no way represent the affections of the different organs, which is to say, the actual diseases.

These groupings of symptoms are derived from abstract entities or beings, entirely factitious, "οντοι"; these entities are unreal, and the treatment one gives them is *ontology* (1821, vol. 2, p. 646).

This passage is cited as a standard criticism of "medical ontology" (Engelhardt, with Erde, 1980, p. 384; Bole, 1995). The criticism is that the nosological frameworks provide specious causal explanations of diseases because the frameworks cannot be manipulated so as to alter the diseased organs. Despite a proposed confusion in grammar (Bole, 1995)[14], Broussais sets the stage for numerous subsequent discussions of a physiological approach to disease (Kraupl-Taylor, 1979, 1983; Rather, 1959; Temkin, 1981; Pagel, 1972; Niebyl, 1971).

Nineteenth-century cellular pathologist Rudolf Virchow (1821-1902) understands cells as defining the seat of disease. The early Virchow understands diseases not as "self-subsistent, self-contained entities" (1981 [1858], p. 188) or invading organisms, but rather as states indicative of underlying pathophysiological processes. Disease represents "only the course of corporeal appearances under changed conditions" (1981 [1858], p. 188). For the early Virchow, disease should not be confused with its causes: "[s]cientific medicine has as its object the discovery of changed conditions, characterizing the sick body or the individual suffering organ. Its object is also the delineation of deviations experienced by the phenomena of life under certain conditions..." (1981 [1858], p. 188). The goal of medicine, then, is to investigate the pathophysiological processes that constitute disease.

The transformation of medicine from a clinical enterprise cataloging clinical entities to a physiologically-grounded one allows the reorganization of the world of symptomatic classifications. Conditions, such as fevers, pain, and diarrhea, are no longer considered diseases in their own right, but symptoms associated with underlying physiological processes that are in turn given a name (e.g., tuberculosis, malaria, Hodgkin's disease). Previously undiscriminated problems (e.g., anemia) could now be distinguished (e.g., general, chronic splenic, and pernicious). Clinical complaints that had not been previously associated (e.g., hepatitis and rabies) could now be brought together under one rubric (e.g., infection). Symptomatically-based names (e.g., phthisis) are replaced with anatomically-based ones (e.g., tuberculosis). These changes lead to the notion that the goal of medicine is found in comprehending pathophysiological processes of clinical problems and take the mere reporting of the symptoms as being less than scientific (Foucault, 1973 [1963]; Engelhardt, 1985, pp. 56-71; Reiser, 1993; King, 1982, pp. 134-137).

The move to understand disease in terms of a clinical model as opposed to syndrome or internal or external substance leads to a variety of ways to understand disease in twentieth century medicine. Cystic fibrosis (CF), for example, is understood in terms of a specific genotype that results in impaired transport of chloride ions across cell membranes, leading to the production of abnormally thick mucus. Thus CF is called a genetic or metabolic disease. Atherosclerosis, which can lead to heart attacks and strokes, may be considered a disease of aging, because it typically becomes a problem later in life after plaques of cholesterol have built up and partially blocked arteries. Measles is considered an infectious disease because it occurs when an individual contracts an outside entity, namely, the measles virus. Schizophrenia is considered a psychiatric disease because it is understood in terms of mental language (e.g., impaired executive functioning, abnormal sensory experiences, and impossible beliefs). In accounting for different relations among clinical phenomena, the world of clinical classification of disease thus expands.

1.2.2. Metaphysical Anti-Realism and The Question of Natural Kinds

From the perspective of nineteenth-century medicine, physiological construals of disease presume for the most part metaphysical anti-realist claims. Metaphysical anti-realism rejects the view that there are real objects that exist independently of our experiences or our knowledge of them. Reality is rather a reflection of our experiences and of the language with which we describe them. In the case of disease, disease is a matter of our ideas or the language with which we name it. In this way, metaphysical anti-realist views of disease are reflected in nominalist (L., *nomen*, name) views.

Lawrie Reznek (1987, 1995) offers a contemporary nominalist defense of disease. After careful analysis, he rejects the view that a natural kind account of disease is available. By natural kind, he means tat which reflects natural order among objects that exist independently of our attempts to classify them. He offers an empirical and an *a priori* argument to support his position. His empirical argument considers the variability of diseases. Some are due to infections, others to deficiencies, and others to structural deformities. In response, he says: "The empirical diversity of the underlying natures...[of health and disease]...counts against us finding some common nature" (1995, p. 575). His *a prior* argument seeks to show that the concept of disease works nominalistically. He counterfactually assumes that a common nature has been found for all diseases (e.g., they are all fungal infections) and then asks whether a condition that would obviously be considered problematic but would not be due to fungal infection (e.g., green spots, fever, etc.) Would count as a disease. Since one knows this would be a disease despite the absence of a common etiology (Campbell et al., 1979), disease is not a natural kind. According to Reznek, "...the disease status of a condition is settled *before* its underlying nature is known. We knew all along that Parkinson's disease was a disease long before the bacillus that caused it was discovered" (1995, p. 575). We knew it was a disease because it is not the underlying nature of a condition that determines whether or not it is a disease; it is the consequences. Given that disease labels turn on consequences (e.g., resulting harm), it follows that there will be conditions that sometimes are considered healthy and sometimes considered harmful, depending on the consequences for the patient, the patient's family, and society.

Take, for example, sickle cell anemia. Sickle cell is the trait of having an abnormal hemoglobin molecule that enables red blood cells to resist malarial parasites. Such a trait is to the advantage of the host in an environment where malaria is endemic. But in environments of low oxygen concentration, such a trait leads to sickle cell anemia. The red cells become sickle-shaped, and rupture, leaving the person anemic. Thus, at high altitudes, the sickle-cell trait would be a pathological condition. Whether it is pathological or not does not depend on something essential in its nature but on the relation among the factors that define the condition. As Reznek concludes, "[t]he existence of continuity between health and disease on the one hand, and health and pathological states on the other, shows that disease and pathological states are not natural kinds" (1995, p. 582; also see Canguilhem, 1978 [1966]).

2. LIMITED REALISM: DISEASE AS RELATION

So far, our analysis has presented two distinct metaphysics of disease in the history of modern Western medicine, One emphasizes disease as a static entity and another emphasizes disease as relation.

Despite the hopes of medical scientists such as Broussais (1981 [1828]), Reznek (1987), and Childs (1999), the move from an ontological to physiological reporting of disease does not eliminate the significance of accounts of signs and symptoms and clinical categories but rather gives such reports a framework for understanding the interrelation between observed phenomena and that which brings such phenomena about. This interweaving enhances the explanatory power of medicine independently of any reference to treatment. With the advent of nineteenth century pathological construals of disease, clinical signs and symptoms are reinterpreted in terms of the basic laboratory sciences, which in turn deliver new clinical descriptions, which are themselves submitted to further revision.

On this view, the interaction between ontological and physiological construals of disease may be understood in terms of the interplay between patient complaints and pathophysiological accounts of clinical problems, which may further be interpreted in terms of the interplay between distinct levels of explanation (*Figure 3*).

Degree of Abstraction	Levels of Explanation
physiological (pathophysiological)	*explanans:* that which explains (methods for knowing)
↕	↕
ontological (patient complaints)	*explanandum:* that which is explained (patient complaints)

Figure 3: *Relation Between the Explanans and the Explanandum*[15]

The interplay occasions a critical dialectic between health care practitioners and medical scientists, about the world of syndromes and the world of pathophysiological correlates, which results in a *limited realist* approach to disease. On the one hand, medical scientists can demand that health care practitioners have their descriptions of clinical problems conform to that which is known about the basic processes of pathology and causal processes. Health care practitioners, on the other hand, can challenge laboratory scientists to provide convincing accounts of what is seen in everyday medical practice. Disease is understood and assessed in terms of an explanatory model, according to the laws of pathology, expressing certain recognized regularities and observed appearances in nature that become the foci of clinical attention.

Each explanatory perspective, each domain of encounter with reality, can criticize and strengthen the others. Where laboratory findings are unable to distinguish, specify, or confirm the presence of particular findings, complaints are labeled mysterious (e.g., chronic fatigue syndrome [Cowley, with Hagar and Joseph, 1990]), or factitious (American Psychiatric Association, 1987, pp. 285-290).[16] Where laboratory data reveal pathology in situations that do not correspond to patient complaints, practitioners' findings are designated as precursors of disease, sub-clinical findings, scientific breakthroughs, lanthanic[17] (Feinstein, 1967, p. 145) or asymptomatic conditions (Billings, 1992, p. 478). When there are neither patient complaints nor laboratory findings, there is no (to the best of our knowledge and technology) clinical problem, including disease. Where both perspectives coincide and support each other (in the case, e.g., of AIDS), there emerges *bona fide* (as opposed to *male fide*) claims of sickness, illness, or disease (Fabrega, 1972; Engelhardt, 1981 [1975]). In short, disease is real, yet bound up with how we know it.

A limited realist view of disease supports the view that disease is a real spatio-temporal object that is in part independent of our experiences or our knowledge of them. Disease has properties and enters into relations independently of the concepts with which we understand them or of the language with which we describe them. Consider that among early hunter/gatherers, infection and nutritional deficiencies were the two most common disease leading to death. In 1900, in the United States, the five most common diseases leading to death were pneumonia, tuberculosis, diarrhea (caused by bacteria and parasites), heart disease, and strokes. In 1988, in the United States, the five most common diseases were heart disease, cancer, strokes, and obstructive lung disease (bronchitis, emphysema, and asthma) (Harrison et al., 1988). Note simply the change in the occurrence of infectious diseases, from their being the primary cause of human disease to their being secondary. Infectious diseases begin to predominate as human communities grow larger and more stable. Increased travel and commerce adds to the spread of infectious organisms from one community to another. But with the availability of antibiotics and accepted hygienic practices in the twentieth century, a marked decrease in infectious disease occurs. One would be hard pressed to say that we simply make infectious disease up. Infectious disease reflects in part a natural reality that has some existence independent of human desires.

Yet, while the evolutionary organic manifestation of disease is real, the language that we bring to it, the interpretations and judgements that we render, are constructed in particular socio-historical contexts. Thus, disease is real, yet humanly constructed. It reflects particular concerns during particular times about human well-being and the desire to avoid pain. In this way, disease is real, yet bound up with human ideas.

3. CLOSING

Ontological and physiological approaches to the metaphysics of disease offer limited yet complementary accounts of disease, thereby resulting in what is called a limited realist approach to disease. There are grounds, then, for holding that disease is in part discovered and in part created, a theme that will be explored in the next chapter, which investigates the epistemological character of disease.

CHAPTER ENDNOTES

1. For discussion on this point, see Po-wah (2002).

2. Here, "ontological" and "physiological" are used in a special sense to indicate adherence to what has been termed "ontological" and "physiological" views of disease (Niebyl, 1971). For a review of the conflicts between "ontological" accounts of disease (those that portray disease as entities, as in some sense things) and "physiological" accounts of disease (those that portray disease taxa as artificial designations), see Engelhardt (1981, [1975], esp. pp. 143-263) and Caplan (1993).

   For historical accounts of the use of "ontological" theories of disease, see Brooks and Cranefield (1959), Cohen (1981 [1961]) and Bole (1995). For philosophical defenses, see Rather (1959).

   For historical accounts of the use of "physiological" theories of disease, see Cohen (1981 [1961]), Temkin (1981), and King (1975). For philosophical defenses, see Engelhardt (1981 [1975]) and Margolis (1976).

3. Much has been written on the nature of paradigm (e.g., Kuhn, 1970 [1962]). I use the term here loosely to refer to a set of scientific and metaphysical beliefs that make up a theoretical framework within which scientific theories can be tested, evaluated, and, if necessary, revised.

4. Nordenfelt (1987, p. 152) makes this helpful distinction.

5. See analyses by Engelhardt (1974), Nordenfelt (1987), Khushf (1995), and Caplan (1997).

6. By reduced, Sydenham does not mean materially sub-divide into the most basic structure of reality, but rather rationally assign to certain categories.

7. Sydenham's appeal to the cause of disease is taken up further in Chapter 5.

8. The distinction between symptoms and signs rests on levels of expression, interpretation, and meaning of clinical phenomena. As King says, "...a symptom is a phenomenon caused by an illness and observable directly in experience. We speak of it as a *manifestation* of illness. When the observer reflects on that phenomenon and uses it as a base for further inferences, then that symptom...is transformed into a sign" (1982, p. 81). Take, for example, fever. Symptomatically speaking, the symptoms involve sensations of heat and perhaps profuse perspiration. The sign for a fever is an observation of redness and wetness and the quantification of heat. Here the sign is what ultimately conveys a clinical message, pointing "beyond itself, perhaps to the present illness [i.e., diagnostic or pathognomonic signs], or to the past [i.e., anamnestic signs] or to the future [i.e., prognostic signs]" (King, 1982, p. 81). In the case of fever, the label points to a present problem, something that has transpired, and something that will occur if the fever is not controlled.

9. A natural kind is a delimited set of genuine properties. At least four senses may be distinguished (Reznek, 1987, pp. 33-48). (1) Whewell's criterion: if the classification reflects an order that we have not merely invented, then there must be a number of logically independent ways of arriving at the same order; (2) cluster criterion: natural kinds tend to cluster; (3) explanatory nature criterion: objects that share a cluster of properties only belong to the same natural kind if they share the same explanatory nature; and (4) law-like connectedness criterion: members belong to natural kinds if they possess a cluster of properties and these properties are connected to one

another in the same way.

10. A discussion of the role values play in disease concepts is taken up in Chapters 6 and 7.

11. Modern physiological views of disease reflect a return to Ancient ways of understanding disease, those particularly developed by the Ancient Greek physician Galen of Pergamun (A.D. 130-201). Galen synthesizes various aspects of Hippocratic medicine and defines the basic features of humoral pathology. Galen speaks mainly of four humors, namely, blood, phlegm, yellow bile, black bile, whereas the Hippocratic authors (Empedocles of Agrigentium and Hippocrates) are not in agreement with respect to their numbers. These humors correlate with the four basic elements (air, fire, water, and earth), with certain natural qualities (hot, dry, wet, and cold), and with specific categories of drugs used to rectify the imbalances (cooling drugs, moisturizing drugs, drying drugs, and warming drugs).

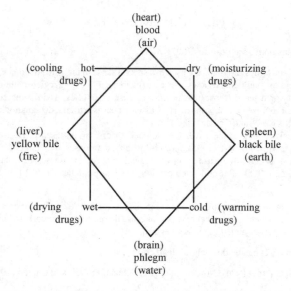

Figure 4: Galen's Theory of Humors

The theory of humors is of great significance to Galen and it forms the basis of his classification of diseases. Fevers, for Galen, are attributed to variations in the balance of humors, which subsequently affect the function of organs. Fevers are due to "the rising heat of decomposing humors," which in turn is the result of the excess or deficiency of one of the four humors (or qualities of hot or cold). Galen believes that there are three principle types of fevers--the phlegmatic, melancholic (black bile), and pirocholic (yellow bile)--corresponding to disorders in the composition of the respective humors. Remedies, such as herbal cathartics and cooling mechanisms, were chosen to rectify the imbalances brought about by fever. Note that this paradigm militates against specific causes for diseases (Eurcrasia). Diseases are defined as mal-mixings and many things could lead to mal-mixings. See W.H.S. Jones (1946), Lund (1936), and Buchanan (1938).

12. This is to underscore a point made by Virchow, but with slight modification, that Morgagni introduced "the anatomical idea" into medical practice. See Virchow (1981 [1895]).

13. Vitalism is the doctrine that there is some feature of living bodies that prevents their nature being entirely explained in physical and chemical terms. This feature may be the presence of a further "thing" (such as a soul), but it also may be simply the emergence of special relations or principles of organization arising from the complexity of the biological organism. The former kind of doctrine understands Life as a kind of fiery fluid (animal, electricity, life force), that needs pouring into an inanimate body in order for it to become alive.

Aristotle (*On the Soul* [1941] and *On the Generation of Animals* [1984]) is a principle source of a more sophisticated vitalism, holding that the life of an animal consists in its *psyche*, which provides a principle of explanation determining the morphological development of the organism, by a principle of teleological or final causation. In the 19th century, the two great exponents of vitalism are Henri Bergson (1859-1941) (1944) and the biologist Hans Driesch (1867-1941) (1914).

　　　　Vitalism has been eclipsed by the advances in molecular genetics, and consequent understanding of the development of organisms in terms drawn from normal science. The consensus among philosophers and biologists is that it offers no explanatory principles that the life sciences need. However, there do remain problems in understanding how different levels of description and explanation, such as those in psychology and in biology, or in biology and in chemistry, relate to each other (e.g., supervenience).

14. Bole (1995) asks how Broussais comes to use *ontoi* for the unreal entities he seeks to criticize. The usage is grammatically deviant, involving an inappropriate plural ending, and the Greek meaning of the term is of things that are real, not ephemeral. Bole suggests that Broussais may have in mind another word, *onkoi*, which has a meaning more in line with his usage.

15. This is indebted to Engelhardt (1982), but expands to include a discussion of clinical models and clinical recipes.

16. Factitious refers to the repeated, conscious simulation of disease for the sole purpose of obtaining medical attention (e.g., Manchausen Syndrome) (American Psychiatric Association, 1987, pp. 285-290).

17. Lanthanic is one whose disease is discovered "accidentally" and has no accompanying complaint.

CHAPTER 4

KNOWING DISEASE

This chapter explores the epistemology of disease, with emphasis on the ways in which concepts of disease reflect how we know[1]. It considers discussions in epistemology about the extent to which reason provides access to knowledge. The focus here is on methods of knowing, on *how* we know, on the *explanans*, in contrast with the previous chapter's attention to *what* is known, on the *explanandum*[2]. This chapter argues that knowing disease requires a *representative realist* method.

1. HOW DO WE KNOW DISEASE?: A DEBATE

Simply put, the debate regarding how we know has been characterized in philosophy as a debate between "rationalist" and "empiricist" approaches.[3] The debate emerges with the assertion by rationalists that reason is a distinct faculty of knowledge, distinguishable from sense experience. With regard to disease, the position is roughly that reason provides access to knowledge about disease. In contrast, empiricists hold that knowledge is derived from sense experience. Empiricists of disease claim that only sense experience provides knowledge of disease. Consideration of the debate between rationalist and empiricist accounts of disease leads to an appreciation of how we know disease.

1.1. Knowledge Through Reason

On one view, clinicians attempt to know timelessly, unconstrained by social and cultural forces. In endeavoring to know truly, clinicians as scientists attempt to understand the world as it would be seen through the Omnipotent's eye, from the viewpoint of a dispassionate, scientific observer, so that findings can be shared with other investigators, even those outside of the scientist's particular culture. New or unclear ideas are evaluated in terms of rational or logical consistency and relation to already established truths. Orderly thinking, or method, ensures that evaluation of ideas properly takes place. The goal here is to attain transcultural accounts of reality. The hope is to attain accounts of disease (e.g., AIDS, tuberculosis, heart disease) that can be diagnosed and known globally. Such is the aspiration of so-called rationalists.

Rationalism is the position that reason has precedence over other ways of acquiring knowledge, or more strongly, that it is the sole path to knowledge. Here reason is understood as the faculty of mind or logical analysis. Rationalism does not generally designate a single precise philosophical position about how reason provides knowledge. To begin with, there are several ways in which reason is understood.

Often, and following Plato (427-347 B.C.) (1992), reason designates a faculty of the soul, distinct from sensation, imagination, and memory, which is the ground of *a priori* knowledge. Yet, there are other conceptions of reason, such as the divinely inspired cause to effect calculation developed by Aquinas (*De trinitate*, qu. 2, a. 3; *Summa Contra Gentiles* I, 4, Baird and Kaufman, 1994, p. 333), the computational account of reason Thomas Hobbes (1588-1679) advances in *Leviathan* (1994 [1651], I.5.), the intuitional account Descartes supports in *Meditations* (1993 [1641]), the narrower account in which Blaise Pascal (1623-1662) opposes reason to "knowledge of the heart" (*Pensées*, 1995 [1670], section 110), and the psycholinguistically-based account Noam Chomsky (1965) provides.

Moreover, there are several ways in which reason can have precedence, and several accounts of knowledge to which it may be opposed. Plato (427-347 B.C.), for example, argues that we have pure intellectual access to the Forms and general principles that govern reality and rejects sensory knowledge of the imperfect realization of those Forms in the material world. Aquinas rejects the subordination of reason to faith and holds that it is impossible for those things revealed to us by God through faith to be opposed to those we can discover by using human reason (*Summa Contra Gentiles*, I, 4, Baird and Kaufman, 1994, p. 333)[4]. Using mathematics as a model, Descartes (1596-1650) argues that all knowledge, including sense knowledge, is deduced[5] from first principles known directly by reason (the *cogito ergo sum* of the *Meditations* [1993 (1641)]). Twentieth century linguist Noam Chomsky (1965) claims that the phenomenon of language requires the postulation of intrinsic intellectual structures or innate mental endowments.

Differences aside, rationalists presume that the attainment of a universal standpoint is possible. New or unclear ideas are evaluated in terms of rational or logical consistency or coherence and relation to already established truths. When one reasons about things, one tries to see the world as reason would see it. By thinking things, we transform them into something universal. Reason as the cement of discourse among persons in general speaks in universals and appeals to the most sustainable generalizations we have of reality. What is sustainable and justifiable in terms of reason reaches from person to person to no particular person. As a consequence, reason's point of view is anonymous, precisely because it is assumable by anyone and belongs to all. It seeks to be ahistorical and unmarked by biases of race, culture, religion, gender, and social status.

1.1.1. Reasoning Disease

Applied to disease, rationalists hold that we discover (L. *dis-* priv., + *cooperire*, to cover, to reveal) disease. We discover disease in the sense that reason reveals or grants access to knowledge of entities that are real and not a figment of our imagination. We discover disease in the sense that pathological processes can be known via the laws of science and medicine. The method allows multiple investigators to arrive at similar, if not the same, conclusions regarding a disease. This is not to suggest that observation plays no role in rationalist thinking. Observation can play a role, but it is never definitive of any certain knowledge. The view is that observation simply cannot establish certain knowledge. Something more is needed. Truth can be evaluated in terms of rational or logical consistency and relation to already established truths. On this coherence view of disease, the word of experts is to be heeded only if it stands the test of rational analysis.

John Brown (1735-1788) (1803) is a rationalist and stands in the tradition of Hoffmann (1660-1742)[6] and Asclepiadean methodism[7]. Brown holds that reason can produce absolutely certain truths about clinical reality, and that some important ultimate truths can be discovered without observation, experiment, or experience.

Brown bases his view of disease on a fundamental principle that he calls "excitability," a basic property of living matter. The correlative principle is excitement, representing the stimuli, either external or internal, that acts on the body. Too much excitement is bad, constituting what Brown calls the "sthenic" diseases. Too little excitement is equally bad, inducing asthenic diseases, which are characterized by disability. A sthenic disease might overstimulate, exhaust the excitability, and thus lead to an indirect debility (King, 1982, p. 232). Diagnosis simply concerns whether a disease is constitutional or local, sthenic or asthenic, and in what degree. Correspondingly, treatment consists in either stimulating or depressing the given condition. To this end, opium (as a stimulant), alcohol (as a depressant), and bleeding (as a depressant) are Brown's preferred therapeutic agents.

For rationalists, and as illustrated by Brown, a major way in which we know disease is by knowing its cause. Here cause is said to be a substantial power or force that brings something about. Three senses of causality [8] are distinguishable in rationalist thought, the first of which represents the ideal to be achieved. First is necessary and sufficient causality. To say that a condition or set of conditions is necessary and sufficient for the occurrence of an event of type E is to say that the following general statement is true: Whenever C occurs, then E occurs, and whenever E does not occur, then C does not occur. Boerhaave (1668-1738), for example, explains disease by causal mechanism related to basic principles such as tonic and atonic conditions. Boerhaave uses the Latin terms *"facit"* and *"producit"* to convey the causal relationship between the altered fibers or humours of the body and the symptoms of disease. For example, whenever *facit* occurs, then disease occurs, and whenever *facit* does not occur, then disease does not occur. Thus, the concepts that Boerhaave uses as the basic explanatory principles is also the cause of disease (King, 1982, pp. 193-194).

To establish laws that express a necessary and sufficient causal relation is for those, such as Boerhaave and Brown, the goal of medicine. Still, many today wish that medicine could provide much better diagnostic data and grant greater certainty regarding the onset and severity of a disease condition. Despite this hope, few examples of necessary and sufficient causal relation are found in medicine because of the inability to list all the conditions that are required for the occurrence of a disease (Albert et al., 1988, Ch. 6). Disease involves more than a single causal variable.

A second sense of casuality, namely, necessary causality, is distinguishable in rational thought. To say that a condition of type C is necessary for the occurrence of an event of type E is to assert the truth of the following generalization: E never occurs in the absence of C. If not E, then not C. The concept of cause as a necessary condition is encountered most frequently in infectious medicine, where a disease-causing role is ascribed to particular agents. In his classic monograph, Robert Koch (1843-1910) (1932 [1882]) establishes the requirements for casual relation between tuberculosis and the bacteria, *tubercle bacilli*. The proof involves the following steps, which provides the basis for what is known as Koch's Postulates: the bacterium must be recovered from the tissues and propagated successively in artificial culture medium until all contaminating material has been removed. When this pure culture, free of all contamination, is injected back into the animal host, there must remain the same disease picture that occurs when the crude material has been injected. If the bacteria in pure culture can produce the same disease as does the crude infectious material, then and only then, can the bacteria be considered the "causal agent" responsible for the infectious property. Following this procedure, Koch establishes that the bacteria is indeed the cause of tuberculosis.

Indeed, other factors in addition to the *tubercle bacilli* are required for the disease to occur. For example, the presence of free oxygen is a necessary condition for cellular metabolism, respiration, and hence for life, including life with tuberculosis. Yet, it would be inaccurate to call the presence of oxygen *the* cause of tuberculosis--or even *a* cause of tuberculosis. Such conditions we typically regard as trivial in an explanatory sense, for they constitute relatively constant background conditions (Mackie, 1965). Usually it is the set of factors that are unusual or that vary from case to case that we examine in order to find the "cause." While many of them can be specified, as stated above, it is not presently possible to list all the conditions that are necessary and sufficient for the occurrence of the disease. Still, tuberculosis is never present unless the tubercle bacillus is also present. Hence, the tubercle bacillus is considered the necessary cause of tuberculosis.

The same can be said of AIDS. According to the National Institute of Allergy and Infectious Diseases (1995), HIV is the cause of AIDS in the sense of being the necessary condition for the disease. AIDS patients have a normal or elevated level of antibodies in their blood, but lack a normal number of helper T-cells or T-4 lymphocytes, which play a crucial role in making antibodies effective. Further, supressor T-cells in CD8 lymphocytes inhibit the antibody system and are present in increased numbers. The immune system is thus severely crippled. This supports the view that a virus-like organism is responsible for attacking the immune system (Gallo and Montagnier, 1987).

A third sense of causality in rationalist thought is sufficient causality. To say that a condition or set of conditions is sufficient for the occurrence of an event of type E is to say that the following general statement holds true: Whenever C occurs (under conditions A), then E occurs. For example, atherosclerosis is sufficient for the occurrence of heart disease. Atherosclerosis is only sufficient because there are other ways to develop heart disease (such as through a genetic condition). In sufficient causal accounts, if we make the causal law specific and if we provide a description, we end up with an account that fits what Hempel and Oppenheim (1948) call a "nomological-deductive explanation,"[9] which may be expressed as follows:

L(1)...L(n) (laws)
I(1)...I(n) (initial conditions)

E (explanandum)

The sentences above the line are known as the *explanans* (that which explains) and the sentence below the line is called the *explanandum* (that which is explained). An explanation is good if (1) the *explanans* is true, (2) the *explanandum* is true, and (3) the *explanandum* is a logical consequence of the *explanans* (Hempel and Oppenheim, 1970 [1948], pp. 10-13). As an example,

Whenever C occurs, E results.
John has C.

Therefore, John has E.

More specifically,

Whenever primary aldosteronism occurs, hypertension results.
John has primary aldosteronism.

Therefore, John has hypertension.

Here primary aldosteronism increases the rate of reabsorption of salt and water in the distal tubules of the kidneys, thereby greatly reducing their rate of excretion. Consequently, mild to moderate hypertension occurs even at normal levels of salt and water intake. Yet, this is not the only way for hypertension to occur. One can have a tumor of the adrenal medulla that secretes large amounts of epinephrine and norepinephrine or be missing a kidney (Goldblatt's hypertension). The point is that primary aldosteronism is at best a sufficient condition for hypertension; there are other causes of hypertension.

For clinical rationalists, then, causal relation plays an important role in knowing disease. Causal relation provides rationalists a conceptual or non-empirical source of knowledge, one that is not reducible to the changing empirical order. As a consequence, rationalists can claim that they provide knowledge that transcends particular socio-historical contexts. While a necessary and sufficient causal relation is least vulnerable to the changing tides of particular socio-historical contexts, rationalists most often have to settle for necessary or sufficient causal relations in their explanations of disease. An exhaustive list of factors contributing to a particular disease type is typically not available and often not needed for diagnostic purposes.

Causal accounts of disease are powerful in that they serve as bases for prediction (L. *prædictio* (-onis), *præ*, before, a foretelling, + *dicere*, to tell). For rationalists, the logical structure of prediction is identical with that of explanation (Toulmin, 1960). The difference is that of the time of occurrence of the event: if the inference is made before the occurrence of the event, the sentence describing the event is a predictive one. If the inference is made after the event has occurred, the sentence is an *explanandum* or explanatory sentence. If explanation consists in knowing the laws and initial conditions, then it is possible to see the event to be explained as fitting into the pattern expressed by the laws. That way, one can be said to understand why that event occurred. By contrast, if we know in advance both the laws and the initial conditions, then we can see that the event will occur. Thus, the same considerations that allow us to understand the event after it has happened would have permitted us to anticipate the event had we possessed the information before its occurrence.

On this analysis, rationalist approaches to disease have much to offer. Rationalist approaches remind us that clinicians do not start off from scratch with every new patient, engaging in new observations every time. Rather, clinicians bring to their observations conceptual constructs (e.g., causal relation) for understanding them that can be shared with others and that create the basis for agreement and disagreement beyond particular empirical orders. The wide range of agreement in clinical medicine concerning categories of disease attest to the role played by prior conceptual claims made possible by reason.

1.1.2. *Distinct But Related Levels of Discussion: Metaphysical Realism and Epistemological Rationalism*
The discussion of epistemological rationalist approaches to disease in this chapter relates to Chapter 3's discussion of metaphysical realist approaches to disease. Where the metaphysical realist claims that there is a real world, the epistemological rationalist claims that reason or a rational method provides access to knowledge of this real world. For the most part, this coupling makes sense, for to claim that there is a world out there entails that the knower can know it via some means that connect or link the knower and the known. John Brown (1735-1788) is an example of a thinker who embraces both commitments in that the material world is simply an indefinite series of variations in the shape, size, and motion of the single, simple, and homogeneous matter that he terms *res extensa* ("extended substance") and reason provides access to such reality.

Yet, metaphysical realist and epistemological rationalist interpretations do not necessarily presuppose one another.[10] One can imagine someone who holds that there are real distinctions in reality, yet rejects the view that reason can know the thing-in-itself. Kant's (1929 [1789]) critical idealism is an example of this position. For Kant, objects of knowing cannot be considered alone themselves, i.e., as "things-in-themselves" (*Dinge an sich*), totally apart from any intrinsic cognitive relation to our representations. Rather, objects of knowing depend on our own faculties of representation, which determine how objects must be, at least when considered just as objects of experience (i.e., of possible human knowledge) or *phenomena*. Knowing objects as phenomena contrasts with knowing them as a *noumena* (i.e., thing-in-themselves), which are specified negatively as unknown and beyond our experience, or positively as knowable in some absolute non-sensible way, which Kant insists is theoretically impossible for sensible beings like us.

Similarly, Sydenham (1981 [1676]) claims that disease structures are real distinctions, but it is questionable whether he thought that reason provides access to things-in-themselves. As the discussion in Chapter 3 illustrates, Sydenham is more concerned with setting forth the basis of an observational method in medicine in the empiricist tradition of his friend John Locke (Romanell, 1974). As he says, the advancement of medicine is contingent on the development of "a history of the disease; in other words, a description that shall be at once graphic and natural" (1981 [1676], p. 146). In describing disease, then, it is necessary "to enumerate the peculiar and constant phenomena apart from the accidental and adventitious ones" (Sydenham, 1981 [1676], p. 147). In short, the relation between metaphysical and epistemological accounts of disease are related but complex.

1.2. Knowledge Through Sense Experience

Empiricism (Gr. *empeiria*, experience) is the epistemological position that sense experience, and not reason or intuition, provides access to knowledge. In endeavoring to know empirically, clinicians as scientists attempt to understand the world as it is, concretely and specifically, from the viewpoint of an embodied and changing knower so that findings can reflect what occurs. New or unclear ideas are evaluated in terms of empirical accuracy and reliability and relation to previously-reported observations. The observational method ensures that evaluation of findings properly takes place. The goal is to attain an accurate and reliable report of the natural order. Such is the aspiration of so-called empiricists.

Empiricism does not generally designate a single precise philosophical position. Classical empiricism, as advanced by Aristotle (*Posterior Analytics* 100a-100b), holds the view that the "universal and necessary" elements of knowledge (which are the foundations of all subsequent reasoning) are built up in the mind through induction.[11] A wider and wider generalization is derived from repeated experiences of particular things until a general or universal concept is established in the mind. Modern empiricism, as espoused by Bacon and Locke (1975 [1690] I, 2, 1), holds that experience is given as a collection of disparate impressions, which are assembled and reassembled by processes of the mind. All ideas are copies of the things that brought about the basic sensations they rest on. Radical empiricism[12] holds that all ideas are reducible to sensations and their relations, as in Hume (1711-1776) (1980 [1739-1740]); nothing else explains knowledge.

In the history of ideas, and for many, the movement from classical empiricism to radical empiricism reflects a decreased confidence in our ability to know. Classical empiricists (e.g., Aristotle) are confident in the ability to arrive at ideas that actually match or correspond with what exists. Modern empiricists (e.g., Hume) are confident in the ability of processing methods to arrive at ideas that somewhat match with what exists. Radical empiricists accept that at best our ideas will match with what is in our minds. The movement from classical to radical empiricism raises concern. If all knowledge comes in the form of one's own ideas, how can one verify the existence of anything external to them? The implication of radical empiricism is for many a skepticism (Gr. *skeptesthai*, to consider or examine) regarding knowledge, followed by solipsism (L. *solus*, alone, + *ipse*, self), which culminates in nihilism (L. *nihil*).[13]

Differences and concerns aside, empiricist epistemology achieves dominance in modern thought in part[14] because of the ascendancy of the physical sciences and the scientific method beginning in the sixteenth century. The scientific achievements of Francis Bacon (1561-1626), Johannes Kepler (1571-1630), and Galileo Galilei (1564-1642) bring about a rejection of speculative inquiries in science and medicine and a reliance on the evidence of one's senses. Among these contributors, special recognition is given to Francis Bacon (1989 [1620]), who inspires Sydenham (1981 [1676]) to talk about method in medicine. In the early seventeenth century, Bacon formulates systematically the elements of what was then called the "experimental method" of natural philosophy, or as we say today, "science." Bacon reacts against a philosophical method (developed by Aristotle in the *Posterior Analytics*) that involves the inspection of ideas on the basis of reason unaided by experience. A principal tool of the method that Bacon rejects is syllogistic reasoning[15], because it is abstract and divorced from experience. For Bacon, science provides the means to study empirically observed events and to identify the cause of those events. Experimental method produces knowledge without appeal to authorities, in particular without appeal to previous published works and the postulations of theologians and public figures.

Bacon's empiricist contribution to the development of science is evident today. Although technology is more complex and scientific knowledge more sophisticated, contemporary scientists still use the requirements for scientific method developed by Bacon. These requirements include careful and rigorous (1) observation, with attention to its reliability, (2) formulation of hypotheses, (3) testing of hypotheses, and (4) induction to general explanation. These requirements continue to provide the bases for the scientific method of investigation today (Biological Sciences Curriculum Study, 1997, p. 21). Scientists seek to establish a connection between a phenomenon and the event purported to be its cause, and this connection to made through a series of steps. These steps include: (1) defining a question based on previous knowledge and rigorous observation, (2) proposing a potential explanation (a hypothesis), and (3) testing the explanation (using experimentation and often statistical analysis). If the evidence supports it, the explanation gains validity. If evidence does not support the explanation, the explanation must be abandoned and a new one sought. Finally, new evidence is tested against existing explanations to ensure that it is adequate. These rigorous steps are necessary in order to distinguish the actual causes from randomly or coincidentally-associated events. Despite the criticisms, empiricism provides a useful basis for the observation method in science and medicine.

1.2.1. Sensing Disease

Modern and contemporary medical professionals rely heavily on sense observations when investigating disease. Early clinicians, such as Sydenham, Sauvages, and Linnaeus, stress the importance of clinical observations at the bedside and describe a number of disease entities on the basis of their clinical manifestations. Nineteenth-century pathologists (e.g., Morgagni) and physiologists (e.g., Broussais) seek to rid medicine of speculative claims and rely solely on empirical observations of anatomical and functional disturbances. Nineteenth-century statisticians such as Pierre-Charles-Alexandre Louis (1787-1872) (1835) and Jules Gavaret (1840) argue that doctors must study as many patients as possible, record data on each, and let the facts, as Auguste Comte (1988 [1830-1842]) would say, speak for themselves.

Despair over the impotence of medicine in a diphtheria epidemic convinces Louis of the necessity of medical statistics in establishing reliable medical knowledge. Louis's principle works are his reports on phthisis (1825), based on 358 dissections and 1960 clinical cases, which point to the frequency of tubercle in the apex or tip of the lung. His statistical work on pneumonia establishes that bloodletting has little value in the condition. His polemics against Broussais (1835) illustrate the strength of the statistical method and the fallacies of an *a priori* theory in medicine. The strong stand that Louis takes in favor of facts and figures and against theorizing of the past does much to advance medical science in the late nineteenth century in Europe and America (Garrison, 1929, pp. 410-411).

For empiricists, a major way in which we know disease is by knowing the general regularity, association, or correlation between events. The claim is that the statement "C causes E" means the phenomenon E succeeds phenomenon C in a regular manner.[16] Take, for example, AIDS. In the early 1980s, clinicians begin to recognize a correlation between the presence of an infection and a certain kind of outcome, one characterized by loss of weight and of immunological protection and recovery. It follows that casually connected events must instantiate a general regularity between like kinds of events. More precisely, if C is a cause of E, there must be types or kinds of events, F and G, such that C is of kind F, E is of kind G, and events of kind F are regularly followed by events of kind G. In the case of AIDS, C (HIV) is seen as a type of an immunodeficiency virus (like HTLV, etc.) and E a kind of response characteristic of loss of immunological function (like HTLV-I and HTLV-II). But HIV (which Gallo et al. [1984] once called HTLV-III) could be distinguished from HTLV-I and HTLV-II in that neither is found in the human species and so must be given a different name. In being open to observation, clinical investigators can expect that knowledge will change and explanations will be revised. Time and time again, clinicians observe that HIV appeared in all individuals with AIDS. The association between the C (HIV) and the E (AIDS) came to be expected. When not all individuals exposed to, or harboring, HIV developed AIDS, clinicians revise their causal claims. Here, the impetus for change or revision is that previous claims have been falsified.[17] An empiricist approach to causality accepts such "new" evidence (in the case of AIDS, evidence regarding immunological protection) and incorporates it into the explanation.

On this view, it is possible to talk about cause and effect, but no appeal to necessary and sufficient, and necessary, causality is made. Rather, empiricists rely on concepts of correlation or association in causal accounts that have nothing to do with the essential nature of the connection between or nature of cause and effect. An event C may be said to be correlated or associated with an event E provided that the association between C and E is commonly observed. Some express the relation in terms of statistical or probability values, where such values are seen to rest not on empirical knowledge but on empirical evidence (Mill, 1874).[18] Statistical laws are statements to the effect that the probability is E that an event of kind C is one that will be followed by kind G. Put schematically, $P(G, C) = E$. Statistical laws that have a form like this are established by collecting data that represent the frequency with which F and G are associated. This accounts for statistical association of events by using statistical methods that reliably distinguish associations owed to chance from those not owed to chance.

In addition to statistical causal relation, empiricists have developed a version of a sufficient causal relation that understands cause in terms of action (Collingwood, 1940, pp. 285-287, 296-311; Beauchamp, 1974). In this version of a casual relation, which may be called a "pragmatic causal relation," the cause is an event or state that we can manipulate to produce another event as an effect. Thus, an event is a cause of another provided that by bringing about the first event, we can bring about the second. In the case of AIDS, for example, C (restrictions on sexual practices) decreases E (AIDS incidences). To ask for the cause of a disease, then, is to ask for the conditions by which humans can prevent it, namely, the conditions that can be controlled in order to prevent the disease from occurring or spreading. Such is the approach used by those who claim that poverty causes AIDS. This account highlights the close connection between cause and action, and the way clinical theory functions as a tool for action, a topic that is taken up at greater length in Chapter 5.

In summary, empiricist views of disease account for the central role sense experience or observation plays in knowing in modern and contemporary medicine. They describe how we synthesize our observations in order to attain knowledge. They account for the evident skepticism in contemporary science and medicine toward speculative assertions in acquiring knowledge.

1.2.2. Distinct But Related Levels of Discussion: Metaphysical Anti-Realism and Epistemological Empiricism

The discussion of empiricist approaches to disease in this chapter relates to the discussion of metaphysical anti-realist approaches to disease considered in Chapter 3. Where the metaphysical anti-realist claims that reality is a set of ideas, the epistemological empiricist claims that sense experience provides access to knowledge. For the most part, this coupling makes sense, as in the case of Louis (1835), for to claim that reality is a set of ideas entails that perception or experience plays a key role in knowing reality.

Yet, metaphysical anti-realist and epistemological empiricist approaches do not necessarily presuppose one another. One can imagine someone holding that we cannot comment on the nature of reality or objects but that we can say something about how we accumulate knowledge of the world. Broussais's (1981 [1828]) empiricism is an example of this approach. In fact, his is a view in contemporary medicine and reflects the prominent one in modern and contemporary medicine that scientists are concerned with natural phenomena rather than ultimate questions about the metaphysical nature of reality. Alternatively, one can imagine someone holding that reality is perception in the mind, but is unable to comment on how we acquire knowledge in any systematic way. George Berkeley's (1685-1753) empiricism is an example of this approach. Berkeley holds that *"esse* is *percipi"* (Berkeley, 1974 [1710], p. 152), "to be is to be perceived," and perceptions "should not have any existence out of the minds or thinking things which perceive them" (Berkeley, 1974 [1710], p. 152). This is not medicine as we know it, but does highlight a growing movement in contemporary medicine to accept a wide range of alternative approaches to healing in medicine insofar as patients believe in their healing effects (Clouser and Hufford, 1993; Cameron, 2000).

2. REPRESENTATIVE REALISM: EXPLAINING DISEASE

So far, our analysis has presented two distinct epistemologies of disease. Clinical rationalism and empiricism offer limited yet complementary approaches to knowing disease. They are limited in that rationalists presume too much (that is, full access to knowing disease without empirical justification) and empiricists risk too much (namely, nihilism).

Alternatively, clinical rationalism and clinical empiricism can be seen to complement one another. Clinical rationalism allows for necessary claims and clinical empiricism frees knowledge from the procedures of mere conceptual analyses. The interaction between rationalist and empiricist construals of disease may be understood in terms of the interplay between clinical reason and clinical observation (Figure 5).

rationalism   ↔   empiricism
(reason)   (observation)

Figure 5: Relation Between Rationalist and Empiricist Approaches to Disease

A critical dialectic is initiated between the world of clinical reason and clinical observation, resulting in what may be called a *representative realist* approach to disease. What further results is a *clinical method*, an orderly way to think about natural phenomena that come to the attention of health care professionals. In this representative realist account of disease, each explanatory perspective, each epistemological approach, can criticize and strengthen the others. Where clinical theory fails to account for the presence of a particular clinical finding, there emerges a category of investigations that is labeled unsupported. In the 1970s, for example, premenstrual syndrome and AIDS could not be explained with supporting data. Where clinical observations reveal pathology in situations that do not correspond to clinical theory, there is an opportunity to reevaluate the observation and revise the theory. Early laboratory work associating the HTLV cell line with HIV (Gallo, 1984) fails to account for the occurrence of an immune deficiency problem in humans. When the observation is accurate and reliable, there is a need to develop a theoretical framework to account for the observation. Here, research on anticoagulants from bat saliva illustrate the advantages of bleeding (Gardell, 1990, 1991). When there is neither theoretical nor observational support, there is no clinical event, including disease. Where both theory and observation coincide, there emerges a *bone fide* focus of clinical investigation, such as found in heart disease.

A representative realist view holds that there is a world of mind-independent objects (e.g., viruses, bacteria, toxins, cells, tissues, organs) that lead us to have experiences. Yet, we do not directly perceive these external objects. What we directly perceive are the effects these objects have on us--an internal image, idea, or impression, a more or less (depending on conditions of observation) accurate representation of the external reality that helps produce it. This subjective, directly apprehended object, has been called by various names: a sensation, percept, sensum, *Vorstellung* ("representation") (Kant, 1929 [1789]). Just as the images on an X-ray represent their remote causes (the structures occurring in the body), the images that occur in the mind, the sense data of which we are directly aware in normal perception, represent their external physical cause. We neither make the world up nor do we know it truly.

A representative realist approach to disease leads us to conclude that disease is in part discovered and in part created. The claim that disease is in part discovered and in part created means that observations in clinical medicine depend on various theoretical presuppositions. Borrowing from Popper, who addresses science and not medicine: "...the belief that we can start with pure observations alone, without anything in the nature of a theory, is absurd" (1965, p. 46). He continues: "Observation is always selective. It needs a chosen object, a definite task, an interest, a point of view, a problem" (1965, p. 46). Alternatively, clinical theory depends on clinical observations or content. Theory without observation is bankrupt; it has no meaning in lived reality. Since "in part discovered and in part created" is a confusing notion in the sense that it sets up a bipolar analysis of knowledge that this work rejects, this work avoids the phraseology. Instead, the term "negotiated," "constructed," "framed," and "reframed" are used to convey the point that in the enterprise of knowing disease, clinical knowers interpret clinical reality (e.g., signs and symptoms, causal agents) through theoretical frameworks that have been submitted to earlier analysis.

3. ON CLINICAL METHOD

A representative realist approach supports a clinical method[19] that relies heavily on observation and first-person reports, explanation and interpretation, and stability and change. This method of clinical investigation is designed to produce an increasingly better understanding of disease. The key here is the adjective clinical (L. *clinicus*, a bedridden person, from Gr, *klinikos*, pertaining to the bed), for the method is grounded in the patient who seeks attention from health care professionals (Ramsey, 1970).

We begin with observation. Observations involve descriptions of events as they occur in the natural world or as the outcome of experiments. Observations in medicine must be as accurate and reliable as possible. Reliable observations are those that are reported by more than one observer and that are in principle replicable by any future observer. Reports that cannot at present be replicated are of very limited reliability, although they might be replicated at some point in the future. Until that occurs, they play only a very limited role in medical investigation. Observations need to be carefully done, so that extraneous factors are ruled out to the greatest extent possible. Observers must report findings precisely. It is one thing to say that the patient's fever is high. It is another thing, and more precise, to say that the patient's temperature is 104 degrees. Observations that are carefully and rigorously obtained and formulated better satisfy the requirements of medical investigation than observations that do not meet this twin test of accuracy and reliability.

Yet, observation is not itself sufficient to understand disease. As a *clinical* science, medicine must study the human entity, in which purpose, values, consciousness, reflection, and self-determination complicate the laws of biology, chemistry, and physics. Medicine must take into consideration the special complexities of the human person as subject interacting with the human person as object of science. Physiology, unlike the clinical science of medicine, studies physical processes, ignoring the lived reality of the experimental subject--his or her self-perceived history, uniqueness, and individuality. Thus, even when it functions as clinical science, medicine must correlate the explanatory modes of the physical sciences with those of the social and behavioral sciences (Pellegrino and Thomasma, 1981, p. 23). Put another way, and in explaining disease, medicine must incorporate first-person or subjective reports of disease into its third-person accounts. It must explain why a patient hurts or suffers. Thus the centrality of the clinical encounter in knowing disease and a new way to understand observation as experiential access to an external world are established (Wartofsky, 1992).

Objective and subjective observations are not sufficient to understand disease. As a clinical *science*, medicine must consider observations from a standpoint of pre-existing knowledge. It brings theory to explain and interpret observations. As a result, a next step is to propose an explanation of third-person observations and first-person reports based on durable knowledge. This step involves hypothesis formation and testing. Based on rigorous observations, medical investigators describe a problem to be resolved. Problems are usually stated in the form of the need for a *causal* account for the observed events. Constant and regular temporal association of events supports the proposal of a causal relationship but clinical medicine requires further analysis. Clinicians use statistical methods to consider large sets of evidence to determine whether their regular and constant association occurs more frequently than chance would predict. When they do occur more often than by chance, clinical investigators say they suspect a causal relation. What results is the possibility of enduring, but dynamic, causal relations, the kind of which will vary in statistical degree of certainty depending on the evidence. In this way, medical investigators test hypotheses first on the basis of durable knowledge and propose explanations based on this existing knowledge. Explanation of disease works hand-in-hand with interpretation. Interpretation links the clinicians with the patient in the clinical context. Two senses of interpretation can be distinguished: interpretation as a method for the historical and human sciences (Dilthey, 2002 [1910]) and interpretation as an "ontological event," an interaction between interpreter and text that is part of the history of what is understood (Heidegger, 1996 [1927]). Providing rules or criteria for understanding what an author or native really means is a typical problem for the first approach. The interpretation of medical evidence provides an example for the first view, since the process of interpreting medical evidence inevitably transforms it. The notion of interpretation adopted in this analysis favors the second position. Interpretation here is understood as that which is part of the finite and situated character of human knowing. It emphasizes understanding as continuing an historical tradition, as well as dialogical openness, in which prejudices are challenges and horizons are broadened (Pellegrino and Thomasma, 1981, p. 23). The causal accounts sought by the natural sciences (e.g., biology) connect sense experience, the discrete representations of outer experience, through hypothetical generalizations. The human sciences aim at understanding (*Verstehen*) that articulate the typical structures of life given the lived experience.

A final step in clinical method occurs when an explanation offered on the basis of existing knowledge is revised or rejected, and then a new explanation is offered. Sometimes it is readily apparent that existing knowledge cannot explain the phenomena in question. For example, some patients who test positive for HIV do not get AIDS. Medical investigators respond to this kind of anomaly by looking to see if anyone has reported similar observations and proposed and proved an explanation of them. Strong statistical association, far beyond that which occurs by chance, is taken to be a mathematical indication of the existence of a causal process. Chance association means that there is at best a temporal regularity in the occurrence of the phenomena. Strong statistical association is also interpreted to mean that other causal processes are probably not contributing to the observed phenomenon.

Transitions from one disease explanation to another will not be merely arbitrary. To recast Lakatos's (1970) concept of research program, one will move from one explanation to another if the second account frees one from the contradictions and ad hoc assumptions integral to the first, allows one to see things more clearly and insightfully, and to approach new problems with greater clarity and insight (Lakatos, 1970). An explanation is strengthened when we have multiple lines of evidence, or evidence from different sources. Different lines of evidence for the same hypothesis reduce the likelihood that we have ignored confounding factors or that we spuriously concluded that our hypothesis predicts the observable data. The more precise our data are and the more precise our hypothesis predicts what we observe, the stronger is our evidence and the more we can say that we know what it is that is being investigated.

On this view, one can expect agreement and disagreement among experts in a particular field. On the one hand, agreement may be widespread. A community of investigators may share common rules of evidence and inference and may acquire sufficient data to explain a clinical problem in a particular way. Much may be agreed upon by all stakeholders. Koch's (1932 [1882]) account of tuberculosis in terms of the *tubercule bacillus* and NIAID's account of AIDS in terms of HIV have received wide support. On the other hand, disagreement can be expected. Participants may share two or more communities of clinical investigators with different appreciations of the rules of evidence and inference, and causal relations, with regard to the issues at stake. One might think here of debates among clinico-scientists regarding the cause of AIDS (e.g., NIAID vs. Duesberg) and the genetic basis of homosexuality (Bayer, 1981; Hamer, 1993).

There are limitations to this process worth emphasizing. Put briefly, it does not lead to final and complete truth. Rather, clinical explanation is open-ended, or open to change. This does not mean that open-ended explanations are false, poorly supported, or not worth paying attention to. Open-ended explanations that result from medical investigation in all cases have greater intellectual weight and authority than proposed explanation that have not been subjected to the requirements of clinical investigation. Medicine may be incomplete, but it is not relative. Disease explanations may be limited, but they are not arbitrary.

4. CLOSING

The analysis so far supports a representative realist approach to disease. Disease is real, yet bound up in an empirical order that we know—we frame according to given and selected epistemological constraints. Put another way, disease is in part discovered and in part created. There is, however, an additional factor involved in the fashioning of disease that has yet to be analyzed, namely, treatment and its role in knowing disease, a topic that is next considered.

CHAPTER ENDNOTES

1. The term "know" (Gr. *gignōsken*, to know) means to perceive with clarity, to understand clearly, or to be sure or well-informed.

2. One is reminded of Marjorie Grene's book, *The Knower and the Known* (1974).

3. This division is in many ways artificial. In his *Lectures on the History of Philosophy* (1955), Hegel rejects a distinction between rationalist and empiricist approaches. I am indebted to Fred Bender, Department of Philosophy, University of Colorado, for our discussions on this and related matters.

4. For more on the relation between faith and reason in Roman Catholic thought, see Cutter (2002e)

5. Here, deduction is the faculty by which subsequent truths are known with necessity from intuited truths, or from intuited truths taken together with other deduced truths (Audi, 1995, pp. 183-184).

6. For Hoffmann, pathology is an aspect of physiology and all of its laws (Garrison, 1929, p. 314).

7. Asclepiades of Bithynia (124 B.C.) establishes Greek medicine on respectable footing in Rome. Asclepiades opposed the Hippocratic idea that morbid conditions are due to a disturbance of the humors of the body (humoralism). He attributes disease to constricted or relaxed conditions of its solid particles (solidism). This is the so-called doctrine of the *strictum et laxum*, which was derived from the atomic theory of Democritus (Garrison, 1929, p. 106).

8. This analysis draws from Albert et al.(1988, pp. 118-125), with the following exception. It interprets "significant" causal accounts in two ways: first, to refer to jointly sufficient causal accounts and second to refer to those accounts in which a factor is chosen as significant in order to achieve a certain goal. Though these two are not incompatible, they are distinct epistemologies. Where the first involves rationalist commitments, the second involves empiricist ones.

9. The "nomological" in the name means that the explanation contains a law or laws. The "deductive" refers to the relationship between the law and the initial conditions and the event to be explained.

10. Here I am indisagreement with Wulff et al., who states that "[t]he justification of realism on the ontological level requires rationalism on the epistemological level" (1986, p. 15). It does not, as this discussion illustrates.

11. An empiricist account of disease employs to some extent an inductive approach to knowing or diagnosing disease. Here induction refers to an inference to a generalization from its instances. Here distinct kinds of inductions may be distinguished. Strict or *explicative* induction is assertion concerning all the entities of a collection on the basis of examination of each and every one of them. The conclusion sums up but does not go beyond the facts. Ordinarily, however, induction is used to mean *ampliative* inference as distinguished from explicative. Ampliative induction is the sort of inference that attempts to reach a conclusion concerning all the members of a class from observation of only some of them. Conclusions inductive in this sense are only probable, in greater or less degree according to the precautions taken in selecting the evidence for them (Audi, 1995, pp. 368-369).

12. James coins the term "radical empiricism." See *Pragmatism and The Meaning of Truth* (1978).

13. This is a complex relation. To begin with, skepticism is the position that knowledge is to be doubted. Some interpret this position to mean that the *individual* cannot be sure of his or her knowledge claims. Solopsists hold that all that the knower can be sure of are the claims that the individual makes. If all one can be sure of are the knowledge claims that one makes, then one might question the possibility of any meaning or knowledge that spans beyond the self. If there is no such meaning or knowledge, a nihilist position results.

14. Not solely, for it is misleading and dogmatic to attribute science strictly to empirical methodology.

15. Syllogistic reasoning is a valid or invalid argument in which a conclusion connecting two terms is deduced from two premises connecting those terms to a third term, called the middle term. The subject of the conclusion is called the minor term, and is connected to the middle term in the minor premise. The predicate of the conclusion, the major term, is connected to the middle term in the major premise, and is conventionally written first. In "All men are mortal, all Greeks are men, so all Greeks are mortal," "men" is the middle term, "Greeks" the minor term, and "mortal" the major term.

    Three main criticisms are made of syllogisms. First, as an argument, they beg the question because the conclusion is already contained in the premises. Second, it is unclear whether the allowed kinds of propositions entail the actual existence of things (called the problem of existential import) (e.g., "all *unicorns* are black"). Third, syllogisms are very limited in scope, and in particular ignore the logic of relations by using "be" as its only verb.

16. A causal relationship is a function of what is sensed, as Hume illustrates. Hume holds that meaningful ideas are analyzable in terms of the sensory impressions from which they are derived. The meaning of the term is exhausted by those features of experience with which the term is in every case associated; that is, meaning is constituted by a set of features, which lead us to say that an item is an X instead of a Y or Z. Armed with this method of proceeding, Hume isolates three empirical relations: contiguity, succession, and constant conjunction. He proclaims them the essential elements of the idea of causation. In considering the action of billard balls, he observed: "In the considering of motion communicated from one ball to another, we could find nothing but contiguity, priority in the cause, and constant conjunction" (Hume, 1938 [1740], p. 22). Additionally, he cites an apparently non-empirical element, necessary connection. "But besides these circumstances, 'tis commonly suppos'd, that there is a necessary connection betwixt the cause and effect..." (Hume, 1938 [1740], p. 22). Hume's theory of causation largely consists of a close analysis of contiguity, succession, constant conjunction, and necessary connection. Special attention is given to constant conjunction and necessary connection, the latter of which Hume thinks to be subjective (or a "habit" of the mind) (Beauchamp and Rosenberg, 1981).

17. Falsification plays an important role in empiricist thinking (Popper, 1959, 1965).

18. Mathematics and logic for Mill (*System of Logic*, 1874) are *empirical* endeavors. Theorems in geometry, for instance, are deduced from premises that are real propositions, inductively deduced.

19. This is not to suggest that there is a single method. Rather, there are variations, as this discussion illustrates.

CHAPTER 5

KNOWING AND TREATING DISEASE

Clinical medicine is an applied science. It is the search for explanation and prediction in the service of practical goals, e.g., the achievement of well-being and the avoidance of impairments. Since one is interested in applying knowledge, the success of such applications is judged primarily by practical standards. Disease concepts involve a complex interplay between two major goals: (1) to know clinical reality, and (2) to alleviate pain and suffering and to prevent premature death and disability. Clinicians wish to establish through concepts of disease the regularities of occurrences among clinical phenomena and to find enlightening and useful models to account for these regularities. In this way, clinicians function as medical scientists seeking to know the world. Ingredient to the task of knowing disease are judgments concerning how to alleviate pain and suffering. As this chapter shows, such judgments are complex and varied. Theory and practice are two aspects of human endeavor that clinicians' intellectual interests sustain.

It is not just that these two endeavors individually and separately influence our understandings of a given disease. There exists, as this chapter shows, a dynamic interplay between knowing and treating, and medical theory and practice. This chapter refers back to Chapters 3 and 4 and extends the conceptual analyses of disease to include the activity of treating (Gr. *therapeia*, from *therapeuein*, to nurse, cure)[1] and a *practical epistemological approach* is offered . It argues that any attempt to develop a purely conceptual or theoretical account of disease is bound for failure. Knowing and treating dually frame what we call disease. As a consequence, the necessary connection between knowing and treating in our endeavors to explain disease becomes evident.

1. MANIPULATING DISEASE: A DEBATE

It is enlightening to construe the shift from the clinical classifications of Sydenham and Sauvages to the anatomical and physiological approaches of Bichat, Broussais, and Virchow as a search for useful ways to manipulate clinical reality, rather than a metaphysical or epistemological search for models and methods. Restated, the shift from syndrome accounts to clinical models of disease may be reinterpreted as the shift from symptomatic treatments to etiological treatments to clinical recipes.

1.1. Symptomatic Treatments: Syndromes as Treatment Warrants

As illustrated in Chapter 3, a prominent way to understand disease in the seventeenth and eighteenth centuries is in terms of syndromes, or constellations of signs and symptoms.   Correspondingly, symptomatic treatments address identifiable constellations of signs and symptoms in order to relieve the complaints patients bring to clinicians. One treats symptomatically because one cannot identify underlying bases of the symptoms. If certain maneuvers act beneficially to change phenomena--to alter or relieve the symptoms, to change or remove the signs--clinicians develop a specific therapy. In this case, theory becomes an instrument for action, and symptomatic treatments result.

Symptomatic treatments are successful insofar as they comfort ailing patients. As seen with AIDS in the early years, palliative treatments (e.g., bed rest, cool baths) provide much needed respite from suffering. Still today, for example, one places a cool wash cloth on the forehead of someone suffering from a debilitating headache (Schmitt, 1987, p. 329). As Chapter 3 and 4 illustrate, clinicians rely heavily on symptomatic reports from their patients, thus carrying out the legacy of Sydenham and the early modern clinicians.

For Sydenham, medicine is to provide a "regular and definite *methodus medendi*" (1981 [1676], p. 152), or set of rules for treating illness, which is to be drawn not from our mere fancies, but rather from experience.

> There must be some fixed, definite, and consummate *methodus medendi*, of which the commonweal may have the advantage. By *fixed, definite*, and *consummate*, I mean a line of practice which has been based and built upon a sufficient number of experiments, and has in that manner been proved competent to the cure of this or that disease.... I require that they be shown to succeed universally, or at least under such and such circumstance (Sydenham, 1981 [1676], p. 150).

The appeal here is to Nature to obtain a clear picture of the rules for therapeutics. Observations, practical experience, and empirical evidence are essential in order to assign empirical truth to therapeutic rules. As Sydenham puts it, "[t]rue medicine consists in the discovery of the real indications [for treatment] rather than in the excogitation [contrivance] of remedies" (Sydenham, 1981 [1676], p. 155).

In establishing a *methodus medendi*, Sydenham appears at times to suggest that one appeals to causal relatedness between an agent and a disease. One is to seek "knowledge of immediate and conjunct causes, things of which the guidance is certain. ...the only causes that can be know to us, and the only ones from which we may draw our indications of treatment" (Sydenham, 1981 [1676], p. 151). For

> ...if he [the physician] know rightly the cause by which it is *immediately* produced, and if he can rightfully discriminate between it and other diseases, he will be as certain to succeed in his attempts at a cure... (Sydenham, 1981 [1676], p. 152).

Here Sydenham's appeal to proximate cause (i.e., that which "immediately" produces an effect) to reveal the proper methods of therapeutics anticipates the thinking regarding causal relatedness of David Hume (1711-1776) (1980 [1739-1740] and late nineteenth-century bacteriologists (e.g., Koch, 1932 [1882]). Sydenham believes that causal relatedness is a matter of direct observation, and that if we can know causes that are sufficiently proximal or immediate, we can have empirical certainty. In contrast, causes that are more remote or more "ultimate" are speculative and not observational (King, 1970, pp. 4-5). In this way, Sydenham appears to support a statistical or probabilistic causal account of disease.

Yet, as Sydenham argues, the discovery of specific remedies need not come from any knowledge of specific causes. "It is not by the knowledge of causes but by that of method...that the cure of the majority of diseases is accomplished" (quoted in King, 1970, p. 5). Careful bedside observation and experience, *not* the discovery of specific causes, or heavy reliance on medical instrumentation (e.g., the microscope), are to deliver knowledge regarding how to treat particular diseases. In the case of a certain disease (e.g., fever), one is to observe signs (e.g., rapid pulse) and symptoms (e.g., reports of feeling hot), and in turn administer therapeutics (e.g., Peruvian bark) that eliminate or mitigate the symptoms (Bowman, 1976a). In this way, an understanding of disease starts from understanding how to reduce patient vexations or complaints on the symptomatic level.

Similarly, Sauvages recognizes the importance of developing nosographies (i.e., clinical descriptions) for useful purposes. He devotes in the last edition of his *Nosologia methodica* (1768) over eighty of the one-thousand-five-hundred pages to etiological and anatomical classifications of disease. Sauvages recognizes that the characterizations of the species of disease should be based on experience with patients and, somewhat pragmatically, that the genera should be given a stable meaning even if such clarity were not fully found in the data. Such an approach is necessary to guide treatment (Sauvages, 1768, Vol. 1, Sec. 73, p. 21).

As seen here, the world of symptomatic treatment focuses on the development of nosographies (i.e., classifications) and nosologies (i.e., descriptions) in order to mitigate, if not eliminate, patient complaints. Prior to the nineteenth century, the clinician traditionally relies on verbal and visual techniques to make a diagnosis. He listens to the patient's description of symptoms and observes the patient's appearance and that of his body fluids. Though ignorant of both the cause of the syndrome and the mode of action of the therapy, the clinician comes into possession of at least a limited explanatory account of the disease, which allows some prediction and control.[2] If certain maneuvers act beneficially to change these phenomena, to treat the symptoms, the clinician gains new understanding of the disease.

1.2. Curative and Preventive Treatments: Etiological Agents as Treatment Warrants

As seen in Chapter 3, with developments in nineteenth and twentieth century science and medicine, disease is no longer understood merely as a syndrome, but as a relation of phenomena understood in terms of pathoanatomical and pathophysiological correlates. The theoretical significance of the shift is justified by its pragmatic force; one would isolate causes that recommend themselves as practicably alterable variables. In the case of AIDS, the isolation of HIV in the mid-1980s becomes an instrument for action or treatment.   In isolating HIV, clinicians come into possession of the conditions that are required for the possibility of contracting AIDS.   Given that explanation serves a predictive function, the search begins for the *magic bullet* that can prevent the onset of AIDS or the other one that can cure it.

Interest in treatment at the etiological level predates nineteenth century medicine. As Chapter 2 indicates, Morgagni (1981 [1761]) and his followers attempt to correlate the pathological lesions found at autopsy with observed symptoms. Leopold Auenbrugger (1722-1809) introduces percussion to detect signs of disease. Microscopical observations carried out by Anthonj van Leeuwenhoek (1632-1723) and Robert Hooke (1635-1703) lead to the discovery of "cells" and ultimately an early account of tuberculosis, then known as consumption (King, 1982, pp. 37ff). Early modern clinicians realize that the manipulation of disease causes may lead to the possibility of manipulating or controlling disease.

But such advancements in disease etiology does not immediately alter the techniques of diagnosis and treatment.   Clinicians resist incorporating etiological accounts into disease explanations in part because they are skeptical toward the reliability of the technology and the replacement of bedside observation with laboratory data. Virchow, for example, is critical of scientific medicine and, of the desire to develop medicine without any appeal to healing sick patients. "According to our view, the concept of medicine, or the art of healing, involves the concept of healing, although according to the newest development of medicine it could appear to have nothing to do with it" (Virchow, 1981 [1847], p. 188). The therapeutic nihilists of the 1840s, Joseph Skoda (1805-1881) and Karl Rokitansky (1804-1878), members of the New Vienna School, rally behind this concern as they criticize medicine for being far more interested in morbid anatomy, or how men die, as opposed to developing curative techniques (Rosenberg, 1979; Akerknecht, 1982, p. 156). Then there is the problem of making sense of therapeutic successes in light of accepted but competing clinical theory.   As an example, the therapeutic success of gold for rheumatoid arthritis makes little sense in light of a vitalist account of disease.

Nevertheless, support for the development of medical therapeutics endures, in part because of emerging success in employing statistical analysis to evaluate treatment. Setting the stage for medical experimentalist Claude Bernard (1813-1878) (1957 [1865]), Jules Gavaret (1840) urges clinicians not to base their therapeutic decisions on speculative theories and logical deductions.   Instead, clinicians must study as many patients as possible and count the numbers of those who die and those who survive. In other words, they must discard theory and rely on positive facts (Tenets #1 and #2, in Wulff et al., 1986, p. 35). Positive facts will, if interpreted carefully, provide "therapeutic laws," or regular associations between observed phenomena. Gavaret reminds clinicians that the accuracy of a statistical law of nature depends on the number of patients and that it is necessary to calculate what statisticians in the late twentieth century call "confidence limits" (Tenet #5, in Wulff et al., 1986, p. 35). In addition, clinicians must distinguish between association by chance and true association and must disregard differences between treatment effects that are not, in twentieth century terminology, statistically significant (Tenets #6 and #7, in Wulff et al., 1986, p. 35).

Guided by developments in medical technology and statistics, major advances occur in late nineteenth and early twentieth century etiological treatments. Etiological treatment seek, on the one hand, to prevent and cure human disease. In terms of prevention, variolation and vaccination, and efforts to control the external environment contribute significantly to the decline of disease. Consider Semmelweiss's (1818-1865) studies of the benefits of washing hands between surgeries, Jenner's (1749-1823) discovery of the milkmaid's immunity to smallpox, Edmund Chadwick's (1800-1890) insight that disease is related to income and wealth, and New York's (Antlers and Fox, 1978) efforts to clean up the city water (Hudson, 1983, Ch. 9). As a consequence of these and related events, clinicians begin to incorporate statistical thinking in their assessments of therapeutic success.

Etiological approaches seek, on the other hand, to cure. Where prevention serves to stop the causal chain of events from even starting, cure (ideally) returns individual to a previous state of health. In the early twentieth century, the development of drugs and surgical interventions contributes significantly to the decline of disease. Highlights include the use of antibiotics in treating bacterial infections (Fleming, 1881-1955), anti-rejection drugs in surgery, and anesthesia in dental work (Wells, 1815-1848) (Hudson, 1983, Ch. 10). In short, etiological approaches to disease remind us that successful treatment methods require more than simply targeting symptoms.

2. CLINICAL RECIPES: CLINICAL MODELS AS TREATMENT WARRANTS

With nineteenth- and early twentieth century developments in clinical therapeutics, there occurs a transition in medicine from a world of symptomatic treatment to one of treating the underlying condition. Within this transition, and despite clinicians' resistance, medicine ventures away from the bedside and moves closer to the laboratory bench, thereby demanding a more sophisticated level of empirical evidence to support its claims.

Yet, despite the emphasis on disease etiology and the search for the magic bullet, medicine remains inextricably tied to the treatment of patient complaints. It must. As Wartofsky says, medicine is a "primary mode of human social practice," a "fundamental mode of cognitive *praxis*" (1976, p. 167)[3], one linked to the lived experiences of those who seek the assistance of health care professionals. Here, *praxis* (Gr. *praxis*, from *prassein*, to do) may be understood to include a whole range of human activities, from what might be regarded more narrowly as technological or instrumental actions (e.g., using medical tools) to the skills, rituals, and forms of human action and modes of human interaction (e.g., creating a patient-physician or patient-institution relationship). In this way, an early meaning of the Greek term *praxis* (doing or action) expands by merging with *poiesis*, or making as in the production of artifacts, arts, and human relationships. In this way, doing and making, and action and production, are compounded, yielding a notion of *praxis* more in line with the understanding developed by Karl Marx (1818-1883) (1961 [1844])[4] in which an intimate relation between modes of social organization and modes of production is proposed (Bernstein, 1971, pp. 11-83; Lobkowicz, 1967).

Medicine as an institution calls for an analysis of the relations among what we do and how we know in order to uncover its clinical strengths as well as its weaknesses. To begin with, in the transition in medicine from a world of symptomatic treatment to one of treating the underlying condition, a critical dialectic is initiated between the health care practitioner and the medical scientist, between that which manipulates (*manipulans*) and that which is manipulated (*manipulandum*) (*Figure 6*).

Levels of Manipulation

manipulans: that which manipulates
(methods for treatment)

↕

manipulandum: that which is manipulated
(patient complaints)

Figure 6: Relation Between the Manipulans and the Manipulandum

On the one hand, symptomatic treatments address individual patient complaints. For example, the provision of bleeding to correct the consequences of a fever targets complaints such as heat flashes and lethargy. Though bleeding may be contraindicated for many reasons, such as loss of blood pressure, it does have its benefits. On the other hand, etiological treatments offer a level of abstraction that can bind syndromes together in terms of underlying causal mechanisms for purposes of manipulation and control. These include, for instance, techniques associated with the goal of the prevention of the underlying problem (through, e.g., behavioral therapies, sanitary reform, vaccination) and the elimination of the underlying problem that results in a cure (through, e.g., drugs or a surgical technique).

The interplay between symptomatic treatment and etiological treatment is dynamic. On the one hand, medical scientists can demand that health care practitioners have their symptomatic remedies conform to what is known about how basic physiological processes may be treated. This is to a certain extent the mission of Donald Seldin's notion that medicine has come "to bring to bear an increasingly powerful conceptual and technical framework for the mitigation of that type of human suffering rooted in biomedical derangement" (1977, p. 40). If one holds, for example, that congenital syphilis spreads from mother to child by a microscopic organism capable of crossing the placental barrier and is highly sensitive to a drug called penicillin, one is not likely to treat the condition with exhortations to the Gods or Goddesses.[5] Health care practitioners, on the other hand, can challenge medical scientists to provide convincing accounts of what is treatable in everyday practice. For those such as Mark Siegler, this clinically-dependent approach to patient care and cure, "how clinical medicine works in the realities of everyday practice" (1981, p.

631), is central to medicine. Specific accounts of disease is understood and assessed in terms of a nomological structure, expressing certain recognized regularities and observed appearances in nature for purposes of alleviating or preventing symptoms and their underlying correlates.

The result of this interplay between symptomatic and etiological treatments is the provision of *clinical recipes* in which structural and theoretical components constituting particular clinical problems are organized in terms of the ways in which select ones can be manipulated and controlled. By allowing the emphasis to fall upon relations rather than upon things, medicine manipulates disease in different ways for different purposes. It develops different modes of intervention (e.g., palliative care, variolation, vaccination, public health measures, drug therapy, surgery, psychotherapy, folk medicine, etc.), depending on the condition under investigation, the availability of means to manipulate it, and the stated goals. Further, it develops the means to coordinate diverse modes of treatment, as in the case of combining a public health campaign with the availability of drugs.

As an example, AIDS can be treated as a genetic, immunological, metabolic, or social condition, depending on whether one is a geneticist, an immunologist, an internist, or a public health official. The construal depends upon the particular observer's appraisal of which variables are most amenable to the manipulation. For example, an immunologist may decide that the major factors in AIDS include immunological dysfunction, which may be altered through immune boosting techniques. The public health official may decide that the basic variables in AIDS are elements of a lifestyle that include risky sexual behavior or use of unclean needles, and ignorance about the consequences, which may be altered by a public health campaign. Further, AIDS can be treated in both ways, thereby bridging distinct therapeutic interventions.

Each practical perspective, each domain of encounter with reality, can criticize and strengthen the others. Where patient complaints are present and unable to be altered, there emerges a category of patient complaints that are labeled "mysterious." Gulf War Syndrome and Severe Acute Respiratory Syndrome (SARS) are examples here. Where following rules for treatment reveal pathology in situations that do not correspond to patient complaints (e.g., BRCA-1), medical findings are designated as sub-clinical findings, precursors of disease, and asymptomatic illness. Advances in genetic testing lead to the isolation of more and more presymptomatic conditions, in which mutations are isolated before they are expressed phenotypically. Where there is neither an object of manipulation nor rules for treatment, there is (at this moment in time) no clinical problem, let alone disease. Where both the object of manipulation and rules for treatment coincide, there is an ideal therapeutic opportunity, such as is the case of treating strep throat with penicillin.

By placing central the manipulative component of medicine, there is an opportunity to reflect on the conditions of *praxis* in medicine, and revise them, if necessary. One challenge comes about because of a change in our understanding of therapy. The more deeply we penetrate this and other situation involving therapy, the more concretely and critically we understand the dynamics of the clinical event, the closer we come to the possibility of transforming medicine into an enterprise that is devoted to responding well to patient complaints. Prior to the nineteenth century, therapy refers to the efficient and productive means to obtain the end of changing a patient's clinical condition. The meaning of therapy changes in the nineteenth century to incorporate both material (physical) and ideal (mental) referents. Therapy refers to a mechanical philosophy but also a systematic division of labor involving a large and complex technological system (e.g., the pharmaceutical industry) in which the boundary between the intricately interlinked artificial and other components (conceptual, institutional, human) are blurred. This blurring of distinctions invites endless reification (abstraction). The concept refers neither to any specifiable institution nor does it evoke any distinct associations of place or of person belonging to any particular group, gender, race, or class. The twentieth century invests therapy with a host of metaphysical properties and potencies, thereby making it seem to be a determinate entity, a disembodied autonomous causal agent of social change. Therapy becomes equivalent to human progress, understood at times to be a "one size fits all" response. This may be cause for concern, as is evidenced by the call for far more individualized therapeutic responses to disease.

The challenge, then, is to resist reifying therapy as the answer to all of our problems. It is to keep the human dimension alive in therapeutic practices. It is to remember that "one size does not fit all" and that things change.

3. INTERPLAY BETWEEN KNOWING AND TREATING DISEASE

The lesson here is not simply that symptomatic and curative treatment interplay, leading to the development of clinical recipes, but that how we know disease depends on how we treat disease, and vice-versa. Knowing and treating are intimately tied in concepts of disease, as illustrated in *Figure 7:*

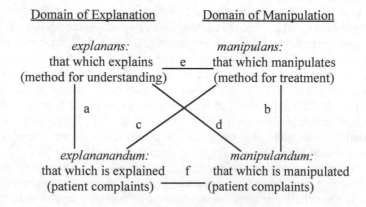

Figure 7: Relation Between Knowing and Treating

With regard to the relation between the world of the medical scientist and the world of health care practitioner, each explains and manipulates the other. Taken together, Chapters 3 and 4 demonstrate the interplay in the domains of explanation and manipulation between the world of the medical scientist and the world of the health care practitioner, (a) between that which explains and that which is explained, by reconstructing the development of disease from an account of a taxonomy of clinical observations to a complex anatomical and physiological reclassification of patients complaints. Correspondingly, Chapter 4 illustrates the interplay in the domain of manipulation between the world of etiologic treatment and syndrome treatment, (b) between that which is manipulated and that which manipulates, by considering the development of clinical therapeutics from a manipulation of patient complaints to a manipulation of underlying conditions (e.g., anatomical, physiological, genetic) associated with the disease. These relations are central to disease concepts. With regard to the relation between that which explains and that which is manipulated (c), the findings of biomedicine function as guides to informing us about what is in need of treatment. An account of AIDS in terms of infectious disease provides the basis for the development of the means to prevent and/or to control the development of AIDS. Conversely, that which is treated calls for an account of *why* it is a legitimate focus of medical intervention. Here conditions such as hyperactivity syndromes (Laufer, 1957; Kaufman, 2001) are examples. The widespread use of ritalin for ADHD fuels the call for a better understanding of why ADHD is a clinical classification.

With regard to the relation between that which is explained and that which manipulates, (d) that which is explained demands the attention of those who develop treatment, and vice-versa. The close connection between what is recognized in the clinic as a disease and the interest of pharmaceutical companies highlights these relations. The development of viagra, for example, is in part a response to the frequency of male impotency seen in the clinic. As a final consideration, (e) that which explains interplays with that which manipulates. Rules for knowing and treating reinforce each other within particular frameworks. For example, a germ theory of disease finds application in efforts directed toward developing the means for therapeutic intervention (Hudson, 1983, p. 193). Alternatively, (f) that which is explained relates to that which is manipulated. This relation is central to early bacteriologists (e.g., Koch, 1932 [1882]; Ehrlich, 1880), for they see themselves knowing and treating the same thing.

This is not to suggest that the relations are always in concert. As the therapeutic nihilists of the 1840s remind us, the relations between the concerns of the medical scientist often diverge from those of the health care practitioner seeking to care and cure patient complaints (Reiser, 1978, 1993; Rosenberg, 1979). In the late twentieth century, the gap between knowing and treating genetic disease has received significant attention, particularly with regard to Huntington disease, for which one can test but cannot treat (Pence, 2000, Ch. 16). Then there is the case of a disease that can be treated (e.g., glaucoma, lupus) but whose mechanisms are still not fully understood. On the one hand, medicine may continue to administer the treatment, despite its harmful or useless effects. Excessive bleeding and the administration of strawberry leaves are examples (King, 1982). On the other hand, medicine may ignore or reject an efficacious treatment for a certain disease because it does not make sense in light of accepted theories of disease mechanism and drug action. In the history of medicine, the rejection of leeching and animal anticoagulants (Gardell, 1990, 1991), gold for arthritis, and aspirin for rheumatoid arthritis (Goodwin and Goodwin, 1984) come to mind. The relation between knowing and treating is central to how we frame disease.

4. TOWARD A PRACTICAL EPISTEMOLOGY

On this analysis, medicine offers a unique opportunity to study the relation between theory and practice[6]. Theories of medicine are themselves instruments of human adaptation and the value of cognitive claims must be judged in terms of their utility, their adaptive value, and their ability to satisfy human needs. In other words, knowing reality is never a purely theoretical endeavor but is a form of practice or *praxis* (Wartofsky, 1976, p. 167). At the heart of this view is a new anthropology that seeks to overcome a dichotomy that plagues modern thought and life: the dichotomy between theory and practice. This is not just an artificial distinction discussed by philosophers. It pervades and shapes our lives. The dominant positivist orientation affects most of our intellectual life and has the consequences of making it difficult, if not impossible, to deal with the vital issues of action and choice that humans inevitably must address in medicine and elsewhere.

Theory and practice are necessarily related. Here, three distinct accounts of the relation between theory and practice may be distinguished (Wartofsky, 1992). A first way to understand the relation between theory and practice is in terms of the relation between knowledge and experience. The relation here is one between the object and the subject, and is set up as one between innatism[7] and empiricism, realism and phenomenalism[8] , or other and the self. As the previous discussion indicates, this relation is not helpful to our analysis because it separates object from subject, and facts or observation from interpretation, which undermines the analysis given thus far (Chapter 4).

A second way to understand the relation between theory and practice is in terms of the relation between theory and observation. The relation here is conceived of as a body of universal statements (in the rationalist tradition) or as a systematically ordered body of data, organized by some functional relation, or by some principle of ordering, such as association or correlation (in the empiricist tradition). The issue that emerges, especially in philosophy of science, is the extent to which our observation, i.e., our experiential access to an external world, is theory-neutral, or the extent to which such observation is theory-laden. On an inductivist-empiricist view, theories are simply abstracted or well-ordered images of observational experience. On an anti-inductivist rationalist view, theories are creations, constructs, imaginative frameworks, or mental entities by which we construe and assimilate experience. This relation is not helpful to our analysis because it separates reason from sense experience, knowledge from practice, and supports a methodology that fails to account for medical knowledge (Chapters 3 and 4).

A third relation between theory and practice is an historical biological one, one that is promoted in this analysis. This relation conceives knowledge as an *active* process, a mode of interaction between an organism and the environment. The contrast here is passive contemplation or reception of experience or merely internal reflex activity by the knowing subject versus active knowing by actual practical activity in the world and as a life-process of adaptation to an environment or the adaptation of an environment to life-needs. This is the sense of theory-practice relation that is most helpful in thinking through epistemological challenges in medicine. Rather than set up a relation between knowledge and experience, or theory and observation, both of which assume passivity in the characterization of the subject's relation to the object of knowing, this relation most closely approximates the formulation of the epistemological challenge of placing central the practical dimension of knowing and the theoretical dimension of acting. Theory and practice cease to be at odds but rather operate *together* in establishing what we know and how we know, practically and historically, in medicine, thus resulting in a *practical epistemological approach*.

The challenge for medicine, then, is to recast the relation between theory and practice in the most concrete and differentiated frameworks provided by specific historical contexts of theory and practice. This is to recognize that "the constitutive activity of the mind, in the relation of theory to practice, takes place...at the fingertips--that is, in the interaction of cognitive organism and environment--in the human case, in the practical activity of intervention in the world" (Wartofsky, 1977, p. 170). The extent to which, for example, our theoretical understanding of our nature (e.g., vitalism, mechanism, dualism) or bodies (e.g., organs, tissues, cells) provides guidance in the development or inhibition of applied medical techniques might be worked out. An understanding of AIDS in terms of an infection guides the development of AZT, ZDV, and timetrexate (Chapter 2). Alternatively, the extent to which the development of medical technology and practice (e.g., the microscope, therapeutic hypnosis, or PCR [polymerase chain reaction]) aids the development of medical explanations may be explored. The widespread use of AZT in the 1990s led to an appreciation of AIDS in terms of a continuum of distinct but overlapping immunological responses.

In addition, the analyses can take other forms, such as investigating the role a social or cultural movement plays in the framing of medical knowledge. Feminism, for example, clearly plays a role in how women's health and disease are framed in the late twentieth century (Chapter 10). In the former U.S.S.R., government concerns about controlling its citizens and avoiding dissension influence how psychiatric labels are fashioned and employed (Pope, 1996). Alternatively, medical knowledge may be seen to influence vast cultural movements, as in the case of the gay rights movement (Shilts, 1987) and the position that AIDS is not a sin of God but a viral condition (Treichler, 1999). Medical genetics and accounts of inheritance are transforming how we understand the global order and relation among cultures. There are, according to Wartofsky, "rich, historical contexts of fundamental and even revolutionary modes of cognitive praxis" (Wartofsky, 1976, p. 188) that await investigation.

In short, medicine offers a study of the relation between knowing and treating, and theory and practice. It offers an example of a practical epistemology that is concerned with how the characteristic human forms of cognitive activity are generated by or constituted by the differential and concrete modes of cognitive practice that emerge in the evolution and history of human life.

5. CLOSING

In this chapter, we come to recognize that medicine seeks not only to know disease, but know it in order to manipulate it, and manipulate it in order to know it. Put another way, medicine without action is theory without practice. Medicine without practice is not medicine as we know it. We are once again reminded of the complex character of disease, a theme that continues with the next chapter's consideration of the role of values in disease.

CHAPTER ENDNOTES

1. There is little analysis of therapy and the relation between knowing and treating in medicine. For discussion of therapy, see Lasagna (1962), Reiser (1978), Rosenberg (1979), Davis (1971), Achterberg (1991), Vogel and Rosenberg (1979). For discussion of medical technology, see Reiser (1978), Davis and Appel (1979), Davis (1971), and Davis (1981). General discussion of the relation between theory and practice is found in Bernstein (1971) and Lobkowicz (1967). Discussion specific to disease may be found in Engelhardt (1986), Pellegrino (1983), Wartofsky (1975, 1975, 1977, 1992), and Hudson (1983, Chs. 9 and 10).

2. In other ways, symptomatic treatments provide little prediction and control. It is difficult to imagine, for instance, that any beneficial somatic effect was obtained by the treatment of a mad dog's bite with angelica root (13th C Denmark), diabetes or piss-pot disease with burnt hedgehog (18th C. England), or gastric ulcer with cantharides (19th C. Denmark) (Wulff, 1981b, p. 123). It was difficult to achieve success absent sufficient knowledge of the course of particular diseases and an appreciation of random variation, the placebo effect, and clinician bias in interpreting results.

3. Also see Wartofsky (1992, pp. 137-138) and Pellegrino and Thomasma, 1981, p. 27).

4. For Marx (1961 [1844]), humans *are* what they *do*; their social *praxis* shapes and is shaped by the complex web of historical institutions and practices within which they function and work. Marx reveals to us and focuses his attention on the "paradox" of human activity in its social forms. For it is the human individual, or rather classes of human individuals, that constantly and continuously create and reinforce the social institutions that pervade human life. These are the objectivications of social *praxis*. And yet these institutions--especially when understood in terms of political economy--have the consequences not of freeing *praxis* and creating those conditions which allow for enjoyment of individuality, but of enslaving or stifling the human individual, dehumanizing and alienating him or her. On this analysis, a goal might be to revise the objectivication of social praxis and thus the alienation of groups of individuals. Again we are reminded of the importance of incorporating the empirical order into our analysis.

5. Some might, of course, as in the case of faith healers (Torkelson, 2000, p. 5A).

6. The view that theory drives practice is expressed in Hegel's account of the theory-practice relation. As he says: "The ultimate aim and business of philosophy is to reconcile thought with reality" (Hegel, 684, tr. Haldane and Simson, 1955, III, p. 545), not to tell the world how it ought to be, and not to try to reform it or to revolutionize it. As a feature of reality, thought or reason is that character of being that permits the mind to

penetrate the whole of reality without presupposing anything except itself. As a human capacity, it is the mind's ability to recognize itself in whatever it approaches. Thought is that which is common to all minds and to all reality, the basic character of what is, the substance and the infinite power of all natural and spiritual life. For Hegel, theory has precedence over practice. "In knowledge alone it knows itself as absolute spirit; and this knowledge, or spirit, is its only true existence" (Hegel, 690, tr. Haldane and Simson, 1955, III, p. 552).

The view that practice precedes theory is supported by Karl Marx. In the history of philosophy, Marx argues that there are "key points" during which philosophy becomes universal and all embracing, thus interrupting its own "linear progress." In the aftertimes, philosophy abandons its purely theoretical attitude and appears on the historical scene as a "practical person (*praktische Person*), hatching intrigues with the world" (Marx, 1961 [1844], 131, as quoted in Lobkowicz, 1967, p. 241). In short, after philosophy has reached a definite degree of universality, it ceases to be contemplative and becomes active. The spectator or philosopher becomes an actor. The philosopher does not become an actor only because the door to further theoretical developments is slammed. Philosophy throws its eyes away "because its heart has become strong enough to create a world" (Marx, 1961 [1844], p. 131, as quoted in Lobkowicz, 1967, p. 241). As Marx puts it, once spirit has reached a definite level of universality, and thus also freedom, it turns into "energy." It becomes will (Marx, 1961 [1844], p. 64, as quoted in Lobkowicz, 1967, p. 241); it becomes action.

7. Innatism (e.g., Ralph Cudworth [1617-1688])) refers to the view that human reason has inherited immutable intellectual, moral, and religious notions, which negate the claims of empiricism.

8. Phenomenalism (e.g., C.I. Lewis [1898-1963], 1946) refers to the study of direct description of our experience as it is in itself without taking into account its psychological origin and its causal explanation.

CHAPTER 6

THE ROLE OF VALUES IN DISEASE

A loud point of contention among scholars who have examined the character of disease is the extent to which value judgments are requisite for or are implicit in these concepts. Perhaps this should come as no surprise given the popular presumption that medicine as a science should be value-neutral because any involvement of values means that the enterprise is subjective. Addressing this are those thinkers who hold that there is no need to resort to value considerations to identify, understand, or analyze accounts of disease. On the other side are those who hold that values play a key role in the identification of diseases. This chapter examines the axiology of disease, and specifically whether values are involved in disease and, if so, what are their nature. It argues that values do play a role and that the values that frame disease are *stipulative* in a *limited* way.

To assess the debate about values, we turn to contemporary scholars. Early modern clinicians do not address this debate in any detail in part because of the taken-for-granted assumption that disease is a value-neutral concept. The quest to cleanse disease of speculative thoughts involved ridding it of values. Much has changed in twentieth century thinking, as the following analysis details.

1. WHETHER VALUES PLAY A ROLE: A DEBATE

At least two positions may be distinguished with regard to the role values play in our understanding of disease. Those who subscribe to value neutralism hold that determinations of disease are matters of empirical fact and not values. Those who subscribe to normativism hold that determinations of disease involve considerations of values, which tell us what is significant and why.

1.1. Neutralist Approaches

Christopher Boorse (1975, 1977) offers one of the clearest expressions of the view that disease can be defined without explicit reference to values. He argues that disease concepts are innocent of essential evaluative components. They are value-neutral and specified in terms of what functions are typical for the individual of the species of which the organism is member. As Boorse frames it:

> 1. A *reference class* is a natural class of organisms of uniform functional design; specifically, an age group of a sex of a species.

> 2. A *normal function* of a part or process within members of the reference class is a statistically typical distribution by it to their individual survival and reproduction....

3. A *disease* is a type of internal state which is either an impairment of normal functional abilities below typical efficiency, or a limitation on functional ability caused by environmental agents.

4. *Health* is the absence of disease. (Boorse, 1977, p. 555)

Boorse understands disease in terms of the capacity to identify derivations from a norm so as to discover what ought to count as disease or dysfunction. Correspondingly, being healthy for Boorse is being a "good specimen" of the species to which one belongs (Boorse, 1975, p. 58). The view here is that the species typical parts or processes of members in a given class of natural organisms delineated by age and sex and contributing to species-survival and reproduction are definitional of "health" (Boorse, 1977, p. 555). It follows that disease is the absence of such qualifications.

On Boorse's view, all organisms, including human beings, are products of a long course of biological evolution. Human evolution has been driven by a wide variety of environmental demands that have conferred advantages on those creatures possessing certain phenotypic and genotypic traits. Since our minds and bodies have evolved in response to our evolutionary past, health consists in our functioning in conformity with our natural design, as determined by natural selection.

Consider, for example, the heart. The heart evolved to perform the function of pumping blood to organs and tissues. When it does, the organism that possesses it can be said to be healthy. If an organism somehow lacks the functions for which it evolved to perform, then the organism is diseased. Values need not enter into the definition of the concepts of disease. This is because evolution is neutral on questions of values. Evolution is neither good nor bad, it just is. It concerns neither benefits nor harms, but rather biological fitness, the implications of which are deemed positive or negative by humans who make judgments.

On this view, disease can be defined as an impairment of the functions typical of a particular biological species, specified in terms of age and sex. These functions are required to achieve the natural goals set, not by politics or culture, but by the twin demands of biological fitness, which include, but are not limited to, survival and reproduction. Survival and reproduction are merely the byproducts of a Darwinian system responding to a specific set of historical circumstances. There is nothing good or bad about them.

Implicit in Boorse's account is a positivist, analytic framework that advances the separability of fact and value, as well as the separability of science, ethics, and metaphysics (the latter of which is often seen as pseudoscience[1]) (Ayer 1935; Stevenson, 1944). In this framework, one has an axiology that sees values as an expression of subjective preferences or emotions, and one advances a "science" that is based on empirical observation and fact, which is objective and thus independent of particular subjects. This view is not new to twentieth-century thought, but rather rooted in modernists' quest to achieve certain and indubitable knowledge, as Chapters 3, 4, and 5 illustrate. This is a philosophical framework that is implicit, as Chapter 3 has shown, in the tension illustrated by the clinic/laboratory dialectic of modern medicine (Khushf, 1995). The ideology of contemporary medicine and science thus supports the particular concept of disease advanced by Boorse and others (e.g., D'Amico, 1995 [2]).

1.2. Normativist Approaches

In contrast, normativist approaches to clinical problems, such as developed by Joseph Margolis (1976), Henrik Wulff (1981b), and H. Tristram Engelhardt, Jr. (1985), argue that disease concepts represent states of the body and mind that are disvalued (e.g., undesirable, useless, bad). The view is that no matter how many descriptive facts are known about the body and mind, it is impossible to decide whether a particular state of affairs represents disease without some reference to values.

Consider a case of tuberculosis (Margolis, 1976, p. 239). To say that Peter has tuberculosis is to say that Peter has a condition that deviates from accepted physiological norms. The concept of a norm is the concept of a condition or parameter in terms of which a range of relevant phenomena may be (evaluationally) graded or ranked as satisfying the given condition. In our case, the presence of tubercle bacillus in the body represents a deviation from accepted norms of physiological functioning. But this is only part of the story. Disease states are in addition evaluative because they typically thwart human goals such as freedom from pain and ability to function and are as such disvalued. Either way, disease involves value judgments concerned with the grading or ranking of phenomena.

Normative accounts explain cross-cultural agreements about certain states being disease (e.g., decreased immunity) on the basis of interference with whatever goals one might have interest in. That is, for many humans, the pain and disability of AIDS will be perceived as dysfunctional and as biologically improper. Alternatively, one can expect that pain (e.g., in childbirth) will not always be dysfunctional. Such value judgments would be made with regard to how certain physiological and/or psychological processes or states that obtain across cultures are experienced in different cultures. The accidentally constant values being given to certain widespread physiologic phenomena (e.g., AIDS, coronary artery disease) and the sensations they evoke (e.g., pain, psychological concern) are the kinds of phenomena normativists rely on to defend their position. In short, allegiance to a normative account of disease does not necessarily commit one to expect great cultural divergence or more seriously the undermining of "scientific" medicine. Rather, one would expect a spectrum ranging from great agreement (e.g., about AIDS) to considerable disagreement (e.g., about culture-bound syndromes) (Bartholomew, 2000; Osborne, 2001).

Put another way for normativists, problems stand out as problems for medicine *because* humans disvalue them. They are seen as pathological. They are associated with pain or suffering, and suffering is judged, for the most part, to have a disvalue. The very appreciation of a problem as a problem for medicine is tied to the patient appearing to have failed to achieve or sustain a desired state of affairs or level of function. Examples include a failure to achieve an expected level of function, an expected realization of human form or grace, or an expected level of freedom from pain or anxiety. Whatever it might be, the very appreciation of a problem as a problem involves the choice to know, to pursue, to fund, and to treat. It depends on a family of values that characterize a circumstance as one of suffering, one of pathology, and a problem to be solved. The problem is of the sort that is beyond immediate willing away, and is embedded in a web of causal forces of anatomical, physiological, and psychological sort open to intervention. Judgments regarding what constitutes disease are inextricably evaluative.

1.3. A Fork in the Road

One is faced, then, with what seems to be a dispute between a single, practically useful conceptualization of disease that reflects what many seek in contemporary medicine, unambiguous answers based on facts, and a view that includes evaluations of the normal and desirable.   Both have appeals. While neutralism appears to grant great stability in clinical nosology and nosography, normativism appears to accommodate the dynamic way that humans organize their judgments about well-being.

Yet, each has its limits. Neutralism fails to account for divergent nosologies across cultures and historical periods. For instance, dyslexia is not a disease in a pre-literate environment, but is in a literate one. Masturbation is a disease in the nineteenth century during a sexually repressive time period (Maudsley, 1868; Engelhardt 1981 [1974]), but not during a time when the prevention of sexually transmitted diseases is of priority. Further, neutralism begs questions on many of the important theoretical issues.   That is, it presumes its conclusion (that disease is factual) in its premises (that statements about disease are factual).   Moreover, it presumes a fact-value distinction that is not defensible by fact or empirical evidence alone.

Alternatively, normativism fails to account for widespread agreement across cultures and historical periods. People do not disagree with the judgments that gout, tuberculosis or leprosy are diseases. Sydenham's (1683) account of gout, and Richard Morton's (1635-1698) (1720) account of tuberculosis (then called consumption) are as relevent today as they were in past times. Further, normativism risks relativism, the view that anything goes. If nosologies simply reflect the values of the classifier, one will not be able to criticize another cultures' classifications. One could not, for example, criticize the Russian practice of institutionalizing dissidents under the umbrella of the psychiatric ill (Pope, 1996; Osborne, 2001, p. 100). Normativism's commitment to difference and change appears to translate into an inability to judge competing positions.

We have arrived at a fork in the road. Before committing ourselves to a position regarding the role of values in disease, let us take a closer look at a related, but under-represented[3], issue, that of the nature of the values that frame disease (Aronowitz, 1998). Upon closer inspection, this issue may be seen to divide those who debate whether values play a role in disease. An elucidation of a geography of views regarding the source of values must assist in resolving the debate between neutralism and normativism.

2. NATURE OF VALUES: A FURTHER DEBATE

The nature of the values that frame disease is a complex topic. The contemporary use of the term "value" (L. *valere*, to be strong) and the discipline now known as axiology are relatively recent developments in philosophy, being largely results of certain 19th and 20th century movements. "Value" is used both as a noun and as a verb. As a noun, it is sometimes abstract and sometimes concrete. As an abstract noun, it designates the property of value or of being valuable. In this sense, "value" is often used as equivalent to "worth" or "goodness." But it is also used more broadly to cover goodness and badness, just as "temperature" is used to cover both heat and cold. In this case, evil is referred to as a negative value and goodness as a positive value. As a concrete noun, singular ("a value") or plural ("values"), the terms refer either to things that have this property of value or to things that are valued. When used as a verb, value denotes a certain mental act or attitude of valuing or valuation.

A distinction is often drawn between two kinds of value, namely, intrinsic value and extrinsic value. By intrinsic value is meant the character of being good or valuable in itself or as an end or for its own sake (Kant, 1985[1785]). By extrinsic value is meant the character of being good or of having value as a means to something (Mill, 2002 [1861]). Here the consequence may refer to actual or potential outcomes. In the end, value is indeed open to interpretation.

The discussion begins here with a general consideration of the sources of the values that frame disease. It addresses whether values are primarily objective or subjective in origin. It is the source question that initially divides those who advocate for the inclusion of values in our understanding of disease.

Among those who think that disease is normative, a distinction is made between those who endorse a version of a value-objectivity thesis (von Wright, 1963; Kass, 1981 [1975]); Pellegrino and Thomasma, 1981, 1988; and Culver and Gert, 1982) and those who support a version of a value-subjectivist thesis (Benedict, 1934a; Szasz, 1961; and Illich, 1976).[4] Those who hold the former subscribe to the view that the values that frame disease are objective or independent of the thinker. Those who subscribe to the latter hold that the values that frame disease are subjective or dependent on the thinker. In short, label this debate as between *value objectivists'* and *value subjectivist's* views of disease.

2.1. Value Objectivism

On the one hand, there are those who hold that the values that frame disease are objective or cognitive[5]. The development of objectivist approaches to accounting for values spans a tradition at least as early as Aristotle (384-322 B.C.) (1985). Aristotle presumes that one could discover the essence of things and that an examination of the essence of man would disclose ingredient purposes or final causes. Following Aristotle, St. Thomas Aquinas (1225?-1274 A.D.) (*Summa Theologica* Q. 91, AA 1-5 [1988, pp. 17-26]) holds that Nature, including man, is created by God and therefore was designed toward the achievement of Divine purposes. Scholars of later periods, such as John Duns Scotus (1265(6)-1308) (1983), synthesize these sets of assumptions regarding the Creator God and His Designs in Nature within an Aristotelian language of essences and final causes. The result is a view that one can discover the essential elements of Nature, including an objective grounding of value judgments regarding ideal states of function, proper form, goals, and standards of right. Appeal to a Designing God makes it plausible to presume that one would thus be able to discover what norms are true, objective, and absolute. Value objectivity affirms that values are independent of individual opinion and ascribes to them a fixed reality common to all.

George von Wright (1963) recasts this Aristotelian view into contemporary philosophical language and address values in disease. Like Boorse, von Wright defines the functions proper to the various organs and faculties as essential functions. But he goes further than Boorse. These functions, he tells us, "are needed for that which could conveniently be called a *normal life* of the individual" (1963, p. 54). The essentiality of the functions does not entail that every individual of the species can actually perform all of these functions. It does entail, however, that if an individual cannot perform a function at the time when by nature it should, the organism is abnormal, defective, faulty, or injured.

A similar line of argument is offered by Leon Kass (1981 [1975]). Kass defines medicine in terms of the ends toward which medicine is directed as a human activity. As he argues, health is

> a natural standard or norm--not a moral norm, not a "value" as opposed to a "fact"...a state
> of being that reveals itself in activity as a standard of bodily excellence or fitness, relative
> to each species, and to some extent to individual, recognizable, if not definable, and to
> some extent attainable (1981 [1975], p. 18).

This Aristotelian analysis portrays health as the well-working of the organism as a whole and an activity of the living body in accordance with its specific excellence. If norms for somatic health and disease can be discovered, the compass of medicine will not depend on individual desires or cultural biases. It will depend instead on what objectively can be known and manipulated. One will be able to know what medicine ought to be doing.

Edmund D. Pellegrino and David C. Thomasma (1981) share with Kass the hope of discovering norms for somatic health and disease. They say, "All cognitive arts share the following characteristics: their theory is a structure of principles about practice; their search is for the right way of making decisions; and their enterprise is intrinsically linked to human purpose" (1981, pp. 177-178). Further they say, "*it is good to be healthy* functions as a moral absolute" (1981, p. 181). Moreover, disease carries certain *universal* characteristics: dependency on others to be cured, ontological assault on personhood, objectivization of the body, dissolution of the self as a unified dynamic construct) (1988, p. 68). The view is that nature delivers concrete values that are intrinsic to medicine. Objective accounts of disease are thus made possible (1988, p. 181).

Charles Culver and Bernard Gert (1982) als provide a normative definition of disease that appeals to objective values. They argue that disease is actually a more general category labeled "malady." The members of this class include not only diseases but also injuries, disabilities, and death itself. They argue that what is common to all these conditions is that human beings *universally* view then as "evils." That is, people disvalue these states and try to avoid them if they can. It is the *perceived* evil associated with dysfunction, and not the dysfunction itself, that is at the heart of understanding the meaning of disease. In the case of AIDS, for example, it is not the deviation from normal functioning that makes us classify this event as a disease. Rather, it is the loss of capacities, the onset of pain, and the risk to life itself, which lead to the disvaluation of this particular deviation from functional normality. While members of different cultural groups or societies may not always agree on what constitutes an evil, every society invariably recognizes certain states of the mind and body (e.g., loss of abilities, loss of freedom, pain, loss of pleasure, and death) as evil to be avoided. As with Pellegrino and Thomasma, objective accounts of disease are made possible.

2.2. *Value Subjectivism*

On the other hand, there are those who hold that the values that frame disease are subjective. The contemporary development of subjectivism spans a tradition beginning with at least the Cyrenaic hedonists (e.g., Aristippus, c. 430-350 B.C., an associate of Socrates). Aristippus holds that we can only know our own sensations but not their objective causes. He draws no inference about the things that affect us, claiming only that external things have a nature that we cannot know. What feels pleasurable or good is pleasure or good. Happiness is the sum of the particular pleasures someone experiences and is preferable only for the particular pleasure that constitutes it.

Aristippus's view has come to be known as *subjectivism*, the belief that we only know our own sensations, and not their objective causes. More specifically, subjectivism is the view that the judgment in question is a report (perhaps a disguised one) about our own attitudes, beliefs, or emotions. Moral judgments, for example, are simply expressions of our positive and negative attitudes (a position called *emotivism* or *noncognitivism*[6] [Ayer, 1935; Stevenson, 1944) or commands (a position called *prescriptivism*[7] [Haré, 1952]), or a matter of convention (a position called *conventionalism*[8] [Poincaré, 1908]).

Consider, for example, pain. Many people experience pain when given an injection. One individual may find a shot mild; another may find a shot excruciating. Are these individuals having the same pain experience but giving them different names? Or are the individuals actually feeling something quite different? Even if both call the same thing "painful," can we be sure that "painful" refers to the same kind of sensation or condition? A subjectivist will respond that all we know is that individuals report their own sensations of pain in different ways. When a patient says that she has a disease, a patient is at best reporting the sensations that define disease for them personally. Correspondingly, clinicians use the term "disease" to report sensations that their patients report. It so happens that similar patients report similar sensations, thus leading to the possibility of classifying disease.

Contemporary subjectivist accounts of the values that frame disease reach popularity in the twentieth century in part because of the success of the logical positivist's project and its influence on discussions regarding the objectivity of values. Ruth Benedict (1887-1948) (1934a, 1934b) views social systems as communities with common beliefs and practices that have become integrated patterns of ideas and practices. Like a work of art, a culture chooses which theme to emphasize from among its repertoire of basic tendencies and then produces a grand design favoring those tendencies. The final systems differ from one another in striking ways, but we have no reason to say that one system is better than another. For Benedict, morality is dependent on the varying histories and environments of different cultures and is no longer directly from the inevitable constitution of human nature (Benedict, 1934b, Ch. 1). Rather, morality differs in every society and is a convenient term for socially approved habits (Benedict, 1934b). Once a society has made the choice of systems and moral codes, normalcy will look different, depending on the select pattern of the culture.

Thomas Szasz (1961) and Ivan Illich (1976) apply a subjective approach to their discussion of disease. Szasz's interest is in the legitimacy of psychiatric labels, but his views are applicable to all disease labels. He claims that mental disease is a "mythological" concept. As he says:

> When we speak of mental illness...we speak metaphorically. When metaphor is mistaken for reality and is used for social purposes, then we have the making of a myth. The concepts of mental health and mental illness are mythological concepts, used strategically to advance some social interests and to retard others (Szasz, 1973, p. 97).

Mental illness advances social purposes insofar as we create our categories of mental illness in order to achieve greater order, or whatever the specific goal is. We single out disruptive kids in a classroom, and label them with "ADHD," for instance, for example, in order to achieve less confusion and more social order for purposes of running (controlling) a class. As a consequences, the concept of mental illness lacks objective medical standing. It justifies its response to mental illness in terms of the achievement of a social good as opposed to the elimination or management of biological dysfunction. For Szasz, then, there are good reasons to question the motivation of psychiatrists who treat patients.

With Szasz, Ivan Illich holds the view that clinical categories are subjective. Yet, Illich takes on the broader thesis that *all* classifications of disease are subjective.

> All disease is a socially created reality....In every society, the classification of disease--the nosology--mirrors social organization...."Learning disability,"

"hyperkinesis," or "minimal brain dysfunction," explains to parents why the children do
not learn, serve as an alibi for schools' intolerance or incompetence; high blood pressure
serves as an alibi for mounting stress, degenerative disease for degenerating social
organization (Illich, 1976, pp. 172-174).

With Szasz, Illich holds that disease classifications reflect nothing more than what
individuals in a culture dislike, find negative, or desire to change. They are socially
created realities and therefore lack any objective standing, a position Sedgwick (1981
[1973]) supports when he says, "[t]here are no illnesses or diseases in nature" (quoted
in Illich, 1981 [1973], p. 121; also see Sedgwick, 1982).

2.3. The Debates Revisited

Given the disagreement on the nature of the values that frame disease, it comes as no
surprise that there is disagreement about the extent to which values frame disease.
Surely, one will not accept an involvement of values in disease concepts with an
unacceptable account of their nature. As a consequence, one can conclude that in
deciding whether values play a role in disease concepts, one must settle the debate
regarding the nature of values. Thus in deciding the role values play in disease, one's
answer to the first debate concerning whether values play a role in how disease is
framed depends on one's answer to the second debate concerning the nature of the
values that frame disease.

3. A LIMITED STIPULATIVE ACCOUNT

Let us return to the debate concerning the nature of the values that frame disease and
consider the objections of each side. What the objectivist finds unacceptable is the view
that the values that frame disease are a matter of personal taste, and that, as a
consequence, any account of disease is acceptable. Disease becomes whatever one
wants it to be. What the subjectivist finds unacceptable is the view that the values that
frame disease are independent of individual interpretation. Subjectivists find no
justification for such metaphysical assertions.

     Perhaps it is possible to resolve the debate by arguing that it is possible to
achieve partial agreement about the property of the values that frame disease, thus
moving the debate beyond a standoff. Consider that objectivists Pellegrino and
Thomasma interpret the values that frame disease as intrinsic and extrinsic
expressions of "organic dysfunction, a perceived need for help, and physician
intervention" (1981, p. 76). Culver and Gert provide a similar analysis. The values
that frame disease express a desire to avoid or eliminate a deviation from functional
normality. Consider that the subjectivist Benedict (1934a) and Illich (1976) interpret
the values that frame disease as subjective expressions of what a culture or individual
dislikes or desires for change. . Notwithstanding their differences, these thinkers share

the view that the criteria for defining disease can be ascertained by *observing* how humans assign worth to lived experiences.

That is, a practical epistemology (Chapter 4) can be used to assess the role values play in disease. For Pellegrino and Thomasma, the test for determining disease involves confirmation of organic dysfunction, perceived need for help, and the existence of physician intervention. For Benedict, the test involves confirmation of socially approved preferences or habits by observation. Medicine allows for this shared sense of empirical value justification. It investigates human as object: as organism that functions and dysfunctions. It is grounded as well in human as subject: as self-conscious, self-perceiving, and self-directing. Both the objective and subjective, observation and first-person reports, explanation and interpretation, stability and change, theory and practice, are necessary for an adequate account of disease (Chapters 4 and 5). In short, it is possible to combine objectivism and subjectivism about values in a broadly practical epistemology, one that takes into consideration various kinds of *concrete experience* as epistemic grounds for beliefs about values.

On this view, a *limited stipulative view* of values results. On this view, values are grading mechanisms employed by limited beings. To begin with, only limited beings grade or make judgments about significance or worth. This is the case because limited beings cannot "have it all"; only unlimited beings are free of boundaries. Limited beings are faced with having to choose or deal with a certain set of options, ones that involve select categories of experience. Humans must choose among or confront options concerning food, shelter, and basic bodily care. Such choices or options are functions of a given biosphere and natural constitution. For humans, these categories or frameworks of thought are ones found in a world comprised of time and space, and are limited in terms of how humans understand and interpret the given world.

In addition, values are matters of stipulation: in arriving at a judgment of value or worth, one must stipulate what it is that is being considered and how it is being considered. Values do not come pre-formed or pre-packaged. Rather, an agent must define the situation, the options before him or her, and the likely consequences (Cowan et al., p. 1992). In this process, agents grade and rank phenomena. In that values are grading mechanisms, they are matters of deliberation and choice. In that there are limits to these choices, the values that frame disease are not open-ended or arbitrary, despite the subjectivist's protestations. In this way, my view parts company with prescriptivist accounts (e.g., Hare, 1952) in that value judgments are not separate and distinct from descriptive judgments. Prescriptivism holds that value judgments serve to command or condemn particular courses of action. They serve as prescriptive or action guides, as opposed to descriptive or factual statements. Their justification lies in a decision of principle or personal choice. While this is in part the case, as we have seen, disease functions as treatment warrants, thus linking description with prescription, and facts with values.

Yet, in that values are treated as distinct yet inseparable from facts, and are not found strictly in the activity or function of objects, this account parts company with a descriptivist one (Foot, 1959; Warnock, 1971). Descriptivism is the view that the meaning of an evaluative term is given without the element of command, or approval, or pressure on action, but is simply thought of as a paraphrase of the natural descriptions of the things to which it applies. The view here differs in that facts in medicine cannot be neutral, but are rather tied to values concerning treating patients (Chapter 4).

A limited stipulative account of values provides what is needed in order to bridge the gap between objectivist and subjectivist accounts of disease. It celebrates what objectivism offers, the stability of the values that frame disease. Yet it provides a way to account for what subjectivism offers, the dynamic and experiential character of weighing and grading reality. Values reflect determinate ways that individual humans grade and rank reality. Values are shared insofar as humans have similar needs (e.g., food, shelter, health care) and interests (e.g., skills, talents, professions). Values differ insofar as humans diverge in their particular or individualized choices or assessments of needs and interests. If this position is acceptable, then it may be possible to define disease as a disvalued biological state of affairs, where such states are defined in terms of *both* the designs and goals of human pychosomatic existence (Chapter 3) *and* first-person reports (Chapter 4).

4. FACTS AND VALUES

At this point, we see that disease involves a complex interplay between two major intellectual goals, (1) to know clinical reality, and (2) to know the significance of human life as it has to do with the alleviation of clinical problems. Clinicians wish to establish through their accounts of disease the regularities of occurrence among clinical phenomena and to find useful and enlightening models to explain these regularities. In this way, clinicians function as scientists seeking to know the world. Ingredient to this task of explaining clinical reality are judgments regarding the goods and worth that define health and about what general accounts show the rationality of such goods and worth and their corresponding duties. In this way, values enter into the way clinicians and patients fashion clinical reality. As Figure 8 illustrates, facts are necessarily tied to values in clinical explanations of disease.

fact   ↔   values

(state of things   (state of things
as they are)   as they ought to be)

Figure 8: Relation Between Facts and Values

It is not just that these two endeavors of knowing and valuing individually and separately influence disease. There exists a dynamic interaction between medical science and value theory, between facts and values. Clinicians, as medico-scientific investigators, organize clinical classifications from a range of measurable and examinable properties of the organism qua organism (Chapters 3 and 4). Clinicians, as interpreters of disease classifications, engage in evaluating findings with respect to clinical outcomes (including, e.g., transaction and opportunity costs) and predictive powers (e.g., usefulness of findings to forecast future events). Laboratory data influences clinicians assessments of the severity of a disease. Value theory provides the general rules for the ways in which we assign significance to, and treatment for, disease (Chapter 5).

The point is that facts and values are necessarily linked in accounts of disease.[9] They are linked because a value-neutral account of disease is usually unavailable. It is unavailable because disease is not only real (Chapters 3 and 4), but bound up with judgments concerning what constitutes appropriate function, human goals, aesthetic expression, and the right thing to do or the right way to be (Chapter 7). It is unavoidable because knowing is a form of doing (Chapter 4), and that it is the full concrete historical person who is the agent of knowledge. On this view, judgments of fact depend on judgments of value.

The fact, for example, that a patient has elevated X (e.g., CD4 T-cells) and decreased Y (e.g., suppress T-cells in CD8 lymphocytes) does not by itself account for disease (e.g., AIDS). The pain and disability of the event, in conjunction with a host of concrete historical and contributing conditions, makes possible the assertion that a patient has a disease. Only by evaluation do we call something such as AIDS "disease" and so see it as part of medicine, as opposed to chemistry or theology.

Put another way, all statements of facts in medicine, however free of evaluation they may appear, are possible only when some act of appraisal has already legislated for the manner of their formulation and assertion. Instead, therefore, of separating judgments of fact from judgments of value as two mutually exclusive classes, I admit both factual and evaluating aspects in all medical judgments. Every medical judgment either includes or presupposes some evaluative component. Some medical judgments, indeed, are more highly evaluative (e.g., psychiatric ones[10]) than others (e.g., cardiac ones), but my point is that even the least evaluative, more "factual," judgments depend for the possibility of their existence on some prior evaluative act.

5. CLOSING

This chapter has addressed the role values play in disease concepts. It argues that values do play a role and that these values are stipulative in a limited way. A limited stipulative account of values lends itself to a discussion of the kinds of values involved in disease because it requires a defense of *how* values are stipulated, a discussion that is next considered.

CHAPTER ENDNOTES

1. The distinction between science and pseudoscience is complex. Briefly put, pseudoscience can be defined as the promotion of unsubstantiated, allegedly scientific opinions. Pseudoscience ideas are built on inaccurate premises, or they do not follow logically from what is observable. Pseudoscience could describe concepts that might be accurate, but the lack of scientific method in pseudoscience assertions prevents us from being able to determine the validity of the ideas. Often, pseudoscience involves claims for which it is almost impossible to provide scientific evidence.

2. D'Amico (1995) shares with Boorse (1975) the view that disease can be understood solely in terms of the causal contributions various states, conditions, and behaviors make to the achievement of their biologically designed ends.

3. Most discussion on the role of values in disease is on the first theme. Minimal discussion (see Mordacci, 1995) takes place on the difference between kinds of values in disease.

4. This division is offered by Mordacci (1995, p. 480). I am in agreement that there are differences in how thinkers understand the source of the values that frame disease but take issue with how certain thinkers are classified. Take, for example, Engelhardt. Mordacci claims that Engelhardt provides the most relativist position of disease. While Engelhardt may be a relativist in terms of the content of values, he is not a relativist when it comes to its source. He holds that certain values (e.g., autonomy) are presuppositions of the moral community and therefore objective. The roots of the position are found in Kant and Hegel. With regard to disease, Engelhardt grants the role of descriptive values in disease. In short, Engelhardt's position on the role values play in disease concepts is more complex that Mordacci indicates.

5. Cognitivism has two expressions, naturalism and intuitionism. For naturalists (Aristotle, 1985; Bentham, 1948 [1789]; Perry, 1954), values are a type of fact, a "natural thing" and value properties are natural properties. Value judgments can be justified through a factual (or "rational") method. Value properties are distinguished from facts and can be ascertained by empirical tests.
    For intuitionists (Plato, 1992; Moore, 1903; Ross, 1930; Ewing, 1947), value terms cannot be defined through factual terms. They refer to nonnatural properties. Value judgments can be true or false because subjects referred to as good or virtuous either do or do not have the value property attributed to them. One "grasps" the truth of a statement. Intuition alone provides confirmation.

6. For non-cognitivists (Ayer, 1935; Stevenson, 1944), value terms do not ascribe properties, and their meaning or use is not descriptive or factual. Rather, the meaning of value terms derives from emotions or attitudes, thereby creating the category of "emotivism." Value judgments are not justifiable through fact, rational method, or intuition.

7. Prescriptivism is the view that value judgments command or condemn particular courses of action. They serve as prescriptive or action guides, as opposed to descriptive or factual statements (Haré, 1952).

8. Conventionalism is the philosophical doctrine that logical truth and mathematical truth are created by our choices, not dictated or imposed on us by the world.

9. Perhaps the best known formulation of the dichotomy between fact and value is put forward by Ogden and Richards (1956), who distinguish between "cognitive" and "emotive" meaning. Correspondingly, the statements of science, which convey information, are distinguished from the "pseudo-statements" of poetry or religious discourse.

10. Reminiscent of the classifications provided by Sauvages (1768) and Cullen (1769), psychiatric classifications are currently based on signs and symptoms. For sure, behavioral genetics, neurobiology, and neuropsychology are moving psychiatry and psychology in the direction of providing etiological accounts,

but this is taking time (National Coalition for Health Professional Education in Genetics, 2002). Here the key is understanding the role genetics plays in cognitive and emotional functioning. In the meantime, syndrome-based accounts lend themselves to value-ladened accounts because they rely heavily on symptomatic reports, and their psychological, cultural, and spiritual influences. This is not much different from how seventeenth century medicine understood somatic disease. Much changes can be expected in the future in psychiatry and psychology.

CHAPTER 7

A GEOGRAPHY OF VALUES IN DISEASE

As Chapters 5 and 6 illustrate, coming to terms with clinical reality is never a purely theoretical endeavor. It is a form of action, which involves diverse modes of intervention and value judgments. In studying the role values play in disease concepts, we are reminded once again that disease is complex. This chapter investigates the ways in which disease reflects what and how we value. Put another way, it investigates the *process* of valuing, as opposed to the previous chapter's emphasis on the *nature* of values. It sets forth a geography of prominent values that frame our understanding of disease.

1. KINDS OF VALUES

The task at this point is to investigate further the values that frame our understanding of disease. As the previous chapter illustrates, the distinction between and among values is diverse and complex. Philosophers from the time of Plato have discussed a variety of questions under such headings as the good, the end, the right, obligation, virtue, moral judgment, aesthetic judgment, the beautiful, truth, and validity.[1] As we see from prior chapters, the values that frame disease are varied and complex. Disease involves states of affairs that are ascribed disvalue, namely, dysfunction, pain, disability, disfigurement, compromised lifestyle, and death, all of which represent less than ideal states of humanness. Further, disease involves states of affairs that are seen to be hindrances to the achievement of certain ends. Dysfunction, for example, hinders the achievement of ability and mobility. Still further, disease involves principles or guidelines that we appeal to as central or indispenable to life. In short, there are at least four kinds of values that frame disease[2]. These include functional (which tell us what ideals of activity are proper to an organism), aesthetic (which tell us what ideals of form and grace are worthy of achievement), instrumental (which tell us how to get from means to end, where the end is the maximization of benefit and minimization of harm, cost, or burden), and ethical (which tell us what ought to be done). The first three are non-moral considerations; the fourth involves ethical or moral ones. Let us consider each in turn.

1.1. Functional Values

It is surely the case that notions of good function surface in talk about effective organ, tissue, or cell action. Here 'function' refers to the normal or characteristic action of something. Since the measure of an organ (e.g., heart), tissue (e.g., cervical), or cell (e.g., pancreatic cell) is highly uninformative, a clinician has no idea from just one single case what range of data can count as "normal" and what range can count as "abnormal." Group data are taken into account. Variations among subjects from different ages and sex must be considered. What is needed is an unambiguous range of measurements for a range of populations that we agree are healthy or diseased. Put another way, disease is to be understood in terms of the concept of malfunction (i.e., a functional ability that falls below other members of the same group at the same stage and of the same sex).

Yet, the search for *the* good functional picture of any particular organ, tissue, or cell is fraught with difficulty. For one thing, the range of measurements is ambiguous. There is rarely a clear cut-off point between good and bad function, between function and malfunction, and between health and disease. Consider atherosclerosis. A cholesterol test finding of 260 mg/dl is no more severe than one of 220 mg/dl, even though the latter is below 240 mg/dl, the threshold that designates atherosclerosis. Much depends on other variables such as the patient symptoms. Similarly, with regard to AIDS, the exact amount of CD4 T-cell concentration that are required for HIV disease will vary (Chapter 2). What is the difference between 180 cells per cubic millimeter of blood and 220, especially when the normal CD4 T-cell levels are at least 800 cells per cubic millimeter of blood? Insofar as medicine is devoted to the care and cure of patients, the move to restrict disease to narrow biological goals on conceptual grounds is not likely to succeed.

Put another way, disease is to be understood in part in terms of dysfunction or impairment, and this is an evaluative notion. Consider the difficulty in identifying physical states as either states of health or disease on a species-typical functional model. Menstruation, for instance, is likely the result of past evolutionary forces that determined when women were best able to bear children. The development of the phenomenon may well confer an advantage in protecting the uterus and oviducts from colonization of pathogens, particularly those brought about by the sperm (Profet, 1993). Yet, one might not want to say that all that accompanies this phenomenon-- painful debilities such as severe abdominal cramping, migraines, loss of blood leading to iron-deficiency anemia--is indicative of health, defined as a regulative ideal of existence (Nordenfelt, 1987). The same could be said of HIV, which may have played a role in protecting humans from the plague (National Institutes of Health et al., 1999b, pp. 6-7). Rather, and despite claims of neutrality, one must recognize that from the human perspective, evolutionary occurrences are sometimes beneficial, sometimes neutral, and sometimes undermining of individual purposes and welfare. Whether something is a sickness or poor training (e.g., stuttering [Preston, 1983]), a defect or decay (e.g., presbyopia), a disease or habit (e.g., alcoholism [Kendell, 1979]), a disease or way of life (e.g., homosexuality [Bergler, 1956; Kinsey, 1953; Feldman and MacCullough, 1971]), or a disorder or a way of life in a highly stimulating environment (e.g., attention deficit hyperactivity disorder (ADHD) [Kaufman, 2001]) turns in part on an account of good or proper function.

The point is that physicians will be unable to determine a classification or treatment of disease by simply attempting to discover what will count as species-typical levels of species-typical functions. Boorse's neutralist view may indeed be of primary interest to biologists engaged in pure or unapplied scientific research. As a zoologist, anthropologist, or paleontologist, one may have an interest in determining what levels of particular function characterize a particular species. One may in addition be interested in discovering the evolutionary genetic processes that lead to those circumstances. But such are not the primary interests of health care professionals or patients who have practical goals such as the relief of pain, the preservation of function, the achievement of a desirable human form and grace, and the postponement of death. It is because medicine has its primary focus on nonremote sets of causes that medicine is not to be equated with evolutionary biology (although clearly, it can benefit from a relation, particularly in genetic medicine [Childs, 1999; Childs and Scriver, 1986; Williams and Neese, 1991; Neese and Williams, 1994, 1998]).[3] At best, Boorse provides a basis for a contrast between evolutionary biology as an endeavor devoted to and medicine as an enterprise devoted to the care of patients.

In addition, Boorse's failure to account for the meaning of disease in terms of function depends in part on a mistaken interpretation of "normal" and "abnormal" as value-neutral concepts. The ambiguity of the term "normal" in medicine is captured well by Edmond Murphy (1976, pp. 123-133). Murphy distinguishes among seven senses of normal: (1) statistical, as with reference to a Gaussian distribution; (2) average or mean; (3) typical or expectable; (4) conducive to the survival of a population; (5) innocuous or harmless, as may be a response by a physician to an inquiry by a patient concerning the significance of particular signs and symptoms; (6) commonly aspired to; and (7) the most perfect or excellent of its class. These seven senses may in turn be clustered around three senses of normal offered by Marjorie Grene (1977, 1978): one is synonymous with "health" (#5 and #6, #7) for Olympic champions; a second is synonymous with "statistically frequent" (#1, #2, #3, perhaps #6); and a third means something like "characteristic for members of the species (#2, #3, #4, and #6)--the usage that Boorse relies on so heavily. The point here is that notions of normal and abnormal, or normal and pathological, vary depending on the goals individuals in different communities wish to obtain. All of this implies what Boorse would call a "normative view of disease" (1975), one that involves values.

1.2. Aesthetic Values

In addition to functional values, aesthetic judgments enter into how we understand disease in the sense that ideal states of physiological, anatomical, and psychological function presume a level of human ability, form, movement, or grace. Such judgments are aesthetic in that form and function are considered beautiful and deformity and dysfunction are considered ugly.

Aesthetics as a branch of philosophy emerged in the eighteenth century in England and on the continent. Recognition of aesthetics coincides with the development of theories of art that group together paintings, poetry, sculpture, music, and dance. Baumgarten coins the term "aesthetics" in his *Reflections on Poetry* (1735) as one of two branches of the study of knowledge, which he argues provides a different type of knowledge from the distinct, abstract ideas studied by "logic." He derives the name from the ancient Greek *aisthanomai* ("to perceive"), and the "aesthetic" has always been intimately connected with sensory experience and the kinds of feelings it arouses. Some have argued that a special perceptiveness or taste is needed to perceive a work's aesthetic qualities (Audi, 1995, p. 10; Ross, 1994).[4]

There are numerous interpretations of aesthetic judgments, but for purposes of this discussion, consider Kant's account as applied to the aesthetic character of the diagnosis of disease (Khushf, 1999). Kant's aesthetics is developed in the context of an analysis of judgment, which is "the ability to think the particular as contained under the universal" (1987 [1890], p. 18). Judgment is divided into two kinds: determinative and reflective. If a universal (concept, principle, rule) is already specified, a particular is brought under it through a determinative judgment. However, if one begins with the particular and seeks out a concept, one that is not yet determined, then reflective judgment is required. Kant separates determinative judgments and science from art, beauty, and reflective judgment.

For Kant, even if a concept (e.g., universal) is not yet available, a particular manifold can be perceived as purposive with respect to the understanding, which as a faculty seeks to bring that manifold into a conceptual unity (what he calls "the objective unity of apperception" [Kant, 1929 (1789), p. 159]. Since the manifold is not yet provided with the needed conceptual unity (the universal is absent), the purposiveness is only *felt*; it is a sense or anticipation that the aim of the understanding will be satisfied (Kant, 1987 [1790], pp. 28-30). This felt but not yet fully realized fulfillment of understanding is a kind of attainment, and "the attainment of an aim is always connected with the feeling of pleasure" (Kant, 1987 [1790]). Kant associates the *feeling of accord* between understanding (with its ends) and the particular, nonconceptualized manifold with *aesthetic pleasure*. Aesthetic judgment is made possible by spontaneous activity in which imagination and understanding are brought into harmony. Through the productive power of the imagination, a nonconceptual unity for the particular manifold is provided, which anticipates a conceptual unity and enables the understanding to come into a purposive relation with that manifold. An object that is perceived in this way as purposeful for the understanding is regarded as beautiful.

Khushf (1999) argues that diagnosis of disease is a form of Kant's reflective judgment. The diagnostic activity can be regarded as an activity that involves both determinative and reflective judgment. The determinative aspect is often emphasized, especially in recent attempts to rationalize medicine. In determinative judgment, specified signs and symptoms are brought under the concept (the disease category) in a relatively mechanical way. However, diagnosis assumes a manifold that is already available for understanding, i.e., one that has been "worked up" so that the relevant information has been isolated and specified. In other words, determinative judgment presupposes an activity in which a given manifold is pre-formed by nature, so that there is an anticipation of the conceptual unity that will be provided by the understanding. That propaedeutic practice involves reflective judgment. The imagination, drawing on the demands of understanding, highlights certain aspects of the manifold. This is a selective process, in which a choice is made among a potentially infinite range of properties. Diagnosis involves a confluence of the preparatory practice of discovery with the broader ideals of medicine. In this way, art constructs medical practice so that it can be a scientific one (Khushf, 1999).

On this analysis, the diagnosis of disease is an aesthetic event, an art practiced by a clinician (Delkeskamp-Hayes and Cutter, 1993). But more importantly for purposes of this analysis, disease itself involves aesthetic judgments. They include ones concerning the form, integrity, harmony, purity, or fittingness of the object. To begin with, disease typically illicits a negative evaluation, namely, one of ugliness or disgust. Recall our discussion of AIDS. In developed countries, such as the United States, individuals wish not to watch wasting people. It's one thing to be skinny, on par with a supermodel, and another thing to be emaciated, especially in a country where food and drink are readily available. In addition, and because of widely available hygienic practices, individuals wish not to witness others with rashes or skin sores. In countries with a distinct Judeo-Christian heritage, many individuals wish not to watch people of the same sex engage in sexual relations. Fundamental assumptions about masculinity (Tobias, 1973, p. 131) are threatened.

There are other examples as well. Tuberculosis patients are quarantined in sanitariums (McKay, 1983) and the possessed (e.g., women) are labelled witches and burned (Achterberg, 1991). Clinicians recommend that women ought not to be educated because learning taxes their system and leads to neurotic behavior, which is unpleasant to watch (Smith-Rosenberg and Rosenberg, 1981). The mentally ill are placed alongside the homeless (Pence, 2000, Ch. 14), and poor body image is treated with human growth hormone, cosmetic surgery, and LASIX. In short, aesthetic judgments contribute to the framing of disease and corresponding interventions.

1.3. Instrumental Values

To continue, the label "disease" serves to place individuals in a therapy role, or to use Talbott Parson's more restrictive term, "sick role" (Parsons, 1958). In determining that someone is sick, one accepts the claim that the person ought to be treated, in that sickness is, most things considered, a disvalued state, and individuals for the most part wish *not* to be in a disvalued state. There is a presumption, then, in favor of such individuals receiving treatment. Correspondingly, clinicians as care-givers seek to alter a state of disvalue in such a way as to allow the achievement of a valued state-- even when this may simply mean an alleviation of the worry of having the disvalued state (e.g., through advice regarding how to accept the new limitations). Central to this endeavor are instrumental values.

Instrumental values guide care-givers in fashioning classifications and rules for treatment. An object has instrumental value if and only if it is a means to, or causally contributes to, something that is desired or valued as a goal. In other words, an instrumental value serves as a means to an end. Concerns regarding achieving desirable consequences such as maximizing benefits (e.g., feeling good, looking good, reducing pain) and minimizing burdens (e.g., pain, harms, high costs) involve instrumental values. Clinicians are concerned to minimize transactions (e.g., financial) and opportunity (e.g., morbidity and mortality) costs to the patient and related parties. If one adopts standards for treatment that are too lax, one may unduly increase the financial, social, and personal costs for care of individuals as well as society at large. However, if one sets the standards for treatment too strictly, one will pay the costs in the loss of lives. As a consequence, one must decide a prudent balancing of the transaction and opportunity costs relating to over- and under-treatment. Such assessments require a prior assessment of the comparative significance of the possible benefits and harms involved in the possible choices.

Consider, for example, cervical cancer (Cutter, 1992). Cervical cancer is the third most frequent malignancy in women in developed countries. Only carcinoma of the breast and the endometrium are more common. About 85% of these neoplasms are squamous- cell carcinomas, and the remaining are adenocarcinomas and mixed adenocarcinomas. The risk of the malignancy is increased in lower socio-economic groups, in those who have early intercourse or have children early, and in those with multiple sexual partners. There are 16,000 new cases diagnosed each year in the United States, 5,000 of which result in death (Rich, 2003).

Figure 9 illustrates the relation between the traditional classification of abnormal "Pap smears" [5] introduced in the 1940s and a new system adopted by medical laboratories in the 1980s. The system incorporates the cervical intraepithelial neoplasm (CIN) terminology to describe various degrees of preneoplastic lesion of the cervical and vaginal epithelium. The CIN system uses three grades of categories, CIN I, II, and III, which correspond to the descriptive terms "mild dysplasia," "moderate dysplasia," and "severe dysplasia" carcinoma *in situ*, respectively.

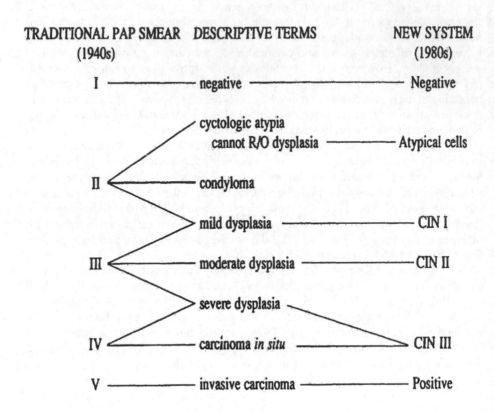

TRADITIONAL PAP SMEAR   DESCRIPTIVE TERMS   NEW SYSTEM
        (1940s)   (1980s)

Figure 9. Classifying the Papanicolaou Smear. Under the system adopted in the 1980s, precursors to squamous cell carcinoma are understood as part of a single disease continuum, what Richart (1976) calls cervical intraepithelial neoplasia (CIN).

The rationale for combining severe dysplasia and carcinoma *in situ* into one category is that (1) severe dysplasia and carcinoma *in situ* cannot be consistently differentiated morphologically, (2) there is no known difference in biological behavior between the two conditions, and (3) women with mild and severe dysplasia tended to be under-treated, and those with small carcinoma *in situ* tended to be over-treated.

          As the third reason indicates, clinicians are concerned to minimize the morbidity, mortality, and financial costs to patients and related parties. In the 1990s, cervical cancer is staged in four, not five, stages (Figure 10).

Clinical Stages of Cancer of the Cervix

Stage I: Cancer confined to the cervix
   IA: Invasive cancer detectable microscopically only
   IA1: Invasion less than 3 mm and width less than 7 mm
   IA2: Invasion more than 3 mm but less than 5 mm
   IB: All others, any visible cancer
   IB1: Cervix less than 4 mm in diameter
   IB2: Cervix greater than 4 mm

Stage II: Spread to adjacent structures
   IIA: Spread into vagina
   IIB: Spread laterally toward the pelvic wall

Stage III: More extensive but still within the pelvis
   IIIA: Extends to the lower vagina
   IIIB: Extends onto the pelvic wall, obstructed ureter

Stage IV: Distant spread or involvement of a pelvic organ
   IVA: Involves the inside of the bladder or rectum
   IVB: Distant metastases, i.e., lung, liver, or bone

Figure 10: Clinical Stages of Cancer of the Cervix (Rich, 2003)

Again, the motivation is to determine the extent of the cancer so that treatment can be more accurately determined. In general, cancer of the cervix are treatment with radiation. The major exceptions are those that are stage I and some that are stage IV. Stage IA cancers that invade less than 3 mm deep can sometimes be treated by simple hysterectomy or even in special cases by cone biopsy. All other stage I cancers are treated either by radical surgery or radical radiation. Some stage IIA cancers can also be considered for surgery. Stage IV cancer requires radical intervention, such as hysterectomy or removal of the pelvic lymph nodes. The goal here is to provide the intervention that brings about the most benefit and least harm to the patient. Careful staging of a cancer according to sound prognostic claims is required.

   Any clinical nosology (e.g., AIDS, coronary artery disease, rheumatoid arthritis) (Gonella et al., 1984) may be seen to take into consideration the extent to which a condition is appropriately treated. Consider three examples involving AIDS. To begin with, the year 1981 marks the emergence of a medical consensus that a pattern of observable signs and symptoms forecasting nearly inevitable death is occurring in isolated groups in the United States (Centers for Disease Control, 1981b; Grmek, 1990)[6]. The Centers for Disease Control begins to receive reports from physicians in New York and Los Angeles about patients with an unusual form of pneumonia caused by the protozoan *pneumocystis carinii*. Signs, such as skin lesions, lymphadenopathy, and unusual infections, and symptoms, such as fatigue, recurrent fevers, unintended weight loss, and uncontrollable diarrhea, are brought together in an unorganized pattern, or syndrome. This syndrome is seen to occur primarily in the "four H groups," namely, Haitians, hemophiliacs, heroine users, and homosexual males. The majority of cases are appearing in the last group, one composed of individuals viewed as highly "promiscuous" (Centers for Disease Control, 1981a). A media frenzy and political activism begin as clinicians attempt to understand the nature of this unusual collection of signs and symptoms (Treichler, 1999).

For a second example, the first natural history study of HIV disease in women was started in 1992 and in 1993 the Centers for Disease Control recognized that HIV-related symptoms specific to women existed. At that time, the agency modified its surveillance definition of AIDS by adding invasive cervical cancer to the list of AIDS-defining conditions, along with pulmonary tuberculosis and recurrent pneumonia (Centers for Disease Control, 1992), thus leading to a marked increase in reports of women infected with AIDS (Faden et al, 1992). In this way, clinicians must decide a prudent balancing of the transaction and opportunity costs relating to over- and under-treatment.

As a third example, in 1996, Owens et al. (1996) proposes a policy for the voluntary screening for HIV in hospital populations where the seroprevalence of HIV is greater than 1%. The authors calculate that counseling and testing costs between $30 and $70 per person with an annual cost of $90-210 million. Such a program would provide for earlier educational and therapeutic interventions for high-risk individuals. For their calculations, the authors presume a 15% reduction in risky sexual behavior and needle sharing. They also presume that 50% of patients would decline testing. Given these assumptions, the authors conclude that such a policy would prevent approximately 565 HIV infections yearly. The results of early intervention, treatment, and counseling would be 50,500 life-years saved in patients and 13,500 life years saved in their partners, with a cost of $36,600 per year of life saved. If calculated in terms of quality adjusted years of life saved, the cost would be $55,500 per year saved. Note how economic forces are instrumental in deciding the boundaries of disease classification.

The point here is that instrumental values shape the very character of disease concepts and place ill individuals in therapy roles. They endow a performative role to our understandings of disease. They are involved in how we classify disease, which is a reflection of what we wish to change and how for purposes of achieving certain goals (e.g., health, time off from work, attention). They highlight the role means (e.g., therapeutics, technology, money, time) play in achieving certain ends. Instrumental values frame our understanding of disease, and tell us what ends are desirable to be achieved.

1.4. Ethical Values

How one understands disease involves choices. One weighs the outcome of a process that has been brought to the attention of a health care professional and the capacity to control the problematic process in a given situation involving particular patients against the harms that may be brought about by not acknowledging the process and not administering treatment. In such cases, there are questions regarding who makes the choice and what choices ought to be made and why (Beauchamp, 1991; Beauchamp and Childress, 2001). Who gets to assign and weigh the values at stake? Who is the moral authority in such situations?

Ethics, or morality,[7] is the study of what constitutes right and wrong, and good and bad, applied to the actions and characters of individuals, communities, institutions, and society. [8] During the last two and a half millennia, Western philosophy has developed a variety of powerful methods and a reliable set of concepts and technical terms for studying the ethical life. Generally speaking, we apply the terms "right" and "good" to those actions and qualities that foster the interest of individuals, communities, institutions, and society. By interest, I mean a share or participation in or a claim. The terms "wrong" and "bad" apply to those actions and qualities that impair interests. But, because trade-offs among moral considerations are complex, constantly changing, and sometimes uncertain, there often are competing, well-reasoned answers to questions about what is right and wrong, and good and bad, regarding complex matters such as how we understand disease.

There are at least four prominent ways to talk about moral considerations. One way is in terms of the results or consequences of actions. [9] Discussion about consequences requires that individuals be able to give well-grounded reasons to explain why an agent or agents should or should not pursue the consequences of an action. Consequences that advance interests are labelled "right" or "good," and an agent or agents should pursue such consequences. Providing medical care to those with conditions that can be addressed by health care professionals (e.g., AIDS patients) is, for the most part, good because this promotes the interests of the parties involved. Consequences that impair interests are "wrong" or "bad" and should not be pursued. Spending money inefficiently and ineffectively on medical care for certain conditions is bad because this impairs the interests of the parties involved by reducing funding for other avenues of established and effective medical care (Weinstein, 1983). With regard to disease, then, we frame disease in terms of the enhancement of good consequences and the avoidance of bad ones, a point made in an earlier section of this chapter.

Another way to talk about moral considerations is in terms of a right or rights. A right is a claim to be treated in a certain way regardless of the consequences of doing so. Recognition of rights expresses the commitment among moral agents to protect and promote self-determination. Members of a free society should pursue protection of rights. Denial of rights can be seen to be a violation of liberty or freedom in a society because denial does not allow an individual to pursue things that he or she chooses. Denial of rights should be prevented in a free society. Alternatively, denial of some rights may be justified when rights are in conflict, or when consequences and rights are in conflict. For instance, there may be grounds for claiming that certain conditions are diseased (e.g., HIV infection [Chapter 2], addiction [Leshner, 1997], social aggression [Brunner et al., 1993]) because it would be unethical *not* to do so from the standpoint of recognizing human rights. One is reminded of concerns raised by the World Health Organization (2003) (e.g., chronic disease and its relation to diet and physical activity, tobacco use and its relation to advertising and promotion). With regard to disease, then, we frame disease in terms of self-determination, not simply because of the expected good consequences but because it is the right thing to do.

A third way to talk about moral considerations is in terms of respect for persons. Brody (1988, pp. 32-33) argues that respect for persons has mistakenly been reduced to a rights account. He points out that there are examples "in which the appeal to respect for persons seems to have a moral significance that cannot be understood along the lines suggested by the rights account" (1988, p. 32). He provides an example of someone selling himself into slavery for food or housing. It is wrong, he asserts, to accept a person's offer to be your slave. In doing so, one would fail to show respect for the individual's personhood, which is different from honoring new type of moral structure. Respect for another, contra Engelhardt (1996), does not mean a minimal recognition of the self-determination of another in the moral community. It does not mean only respect for rational agents. To show respect for persons is to value persons by *refraining* from eliminating the necessary conditions of personhood, which include life, bodily integrity, freedom to make choices and to act upon them, and so on. Correspondingly, it means *acting to promote* the presence of such conditions. Respect involves, then, a negative and positive duty to others. On this view, respect is not dependent on the consent or rights of another. The obligation to show respect for persons is not an obligation to the person in question. It is an obligation to act in a certain way toward that person or persons in order that the necessary conditions of personhood may be realized. And so, respect is owed to the innocent and vulnerable (e.g., children), to communities of persons (e.g., particular Native tribes [Freeman, 1997; Stapleton, 1995]), as well as so-called rational agents (Engelhardt, 1996, Ch. 2). With regard to disease, then, we frame disease in terms of how we promote respect for persons. All things considered, we ought to treat persons who are diseased (e.g., with HIV infection) if something can be done.

A fourth way to talk about moral considerations is in terms of virtue, or moral character. Plato (1992) and Aristotle (1985) did not discuss what actions will lead to the best consequences, what rights people have, or what respect people are owed. These ancient Greek philosophers discussed what sort of person one ought to be and what sort of actions one ought to perform as part of being the right sort of person. Despite its decreased popularity today[10], a morality of virtues is in fact commonly appealed to in medicine. As Brody (1988, pp. 35-42) notes, many decisions made by health-care providers and patients are justified at least in part on the grounds that only by making these decisions can they maintain particular character traits that are considered laudable, such as integrity, compassion, courage, and beauty. Since such decisions occur in the clinical setting, one might ask how decisions about disease involve appeals to virtue. In the development of disease classifications, we often can detect an appeal to a standard of character, as in the case of AIDS. In the early years, AIDS was considered by many to be a breakdown in moral character, particularly the ideals of sexual prudence and temperance[11]. In addition, one might wonder about the extent to which concerns about beauty influence clinical classifications in dermatology. Such influence contributes to the development of disease classification and the search for treatment.

2. CLOSING

Knowledge of disease is not simply for knowledge's sake, it is knowledge for the sake of action. Disease exists in order to cure, to care, and to intervene. Since we understand disease in terms of responding to patient complaints, it follows that all disease takes place against a background of value presuppositions. Functional values tell us what ideals of activity are proper to an organism. Aesthetic values tell us what ideals of form and grace are worthy of achievement. Instrumental values tell us how to get from means to end, where the end is the maximization of benefit and minimization of harm. Since all may not share in the same views regarding what ideals are proper to be achieved, ethical values are at stake as well. The determination of disease is in the ultimate analysis the result of value choices. To diagnose, predict, and treat, while taking proper regard of patients, requires acknowledging the range of values embodied in disease concepts.

At this point, we recognize the futility of discovering *the* essential structure of disease and of developing an understanding of disease divorced from patient complaints, action, and values. Absent evaluative concerns, disease concepts would have no stake in benefitting patients, in seeking good consequences, and in warranting certain actions. Theory without action in medicine is theory without practice. Medicine without practice is not medicine as we know it. It follows, then, that the contexts of disease are critical in understanding disease, a topic that is next considered.

CHAPTER ENDNOTES

1. In the nineteenth century (Meinong, 1894), the conception was born that all these questions belong to the same family since they all are concerned with value or what ought to be.

2. There are a variety of suggestions regarding how values in disease are to be distinguished. Aronowitz (1998), in particular, does a laudable job diagnosing the force of values in specific disease types (e.g., chronic fatigue syndrome [CFS], ulcerative colitis, Lyme disease, and coronary artery disease). His first case study, CFS, serves as a template for others that are less obviously value-laden. Aronowitz notes six nonbiological values that shape the story of CFS. These are public and professional attitudes and beliefs, professional status, media coverage, ecological relationships, therapeutic and diagnostic practices, and economic conditions. Aronowitz's case-oriented approach to understanding how values shape disease highlights the varied and concrete nature of the values that frame specific disease classifications. Although this is an approach that has much merit (Erde, 2000, p. 590ff), it underestimates the shared way that humans value. The analysis in this chapter offers such a generalization, which is not meant to contradict a case-study approach. One might note that instrumental values (which tell us how to get from means to ends) are at play in clinicians' assessments of media's influence on the development of clinical nosologies and of economic benefits and costs. Alternatively, clinicians' assessments of the permissible boundaries of therapeutic and diagnostic practices entail judgments about what ideals of activity are proper to an organism. In other words, functional values are at play in disease.

3. One benefit is understanding the role disease has played in the evolution of the species (Childs, 1999; Williams and Neese, 1991; Neese and Williams, 1994, 1998). One might be familiar with the evolutionary advantages of sickle cell as a prevention against malaria and clinical depression as a way to deal with stressful times. Traits that help organisms to produce children are passed on to future generations.

4. The roots of this view may be found in Socrates, and his discussion of the educability of taste.

5. The Paps smear is named after Dr. George Papanicolaou, who developed the test in the 1950s.

6. Some scientists believe HIV spread from monkeys to humans between 1926 and 1946 and first appears in Africa in the 1930s.   Others claims that, in 1959, a man died in the Congo of what we now call AIDS. Others report that gay men in the U.S. and Sweden–and heterosexuals in Tanzania and Haiti–begin showing signs of what we now call AIDS. In *And The Band Played On*, journalist Randy Shilts designates Gaetan Dugas (at the 1980 San Francisco gay pride parade) as "Patient Zero," the man whose erotic penchants and compulsion put him causally at ground zero of the American AIDS epidemic (1987, p. 11).

7. Ethics is the study of morality, i.e., standards or judgments concerning right or wrong, and good and bad, conduct or behavior. The word is commonly used interchangeable with "morality" to mean the subject matter of this study; sometimes it is used more narrowly to mean the moral principles of a particular tradition, group, or individual.   The term "ethical" is from the Gr. *ēthikē (technē)*, ethical or art. "Moral" is from the L. *moralis*, moral, pertaining to manner, from *mos*, manner.

8. Larry McCullough and I developed this analysis in Biological Sciences Curriculum Study (1992).

9. This discussion of moral considerations in terms of consequences shares important features with our discussion of non-moral considerations in terms of instrumental values. In particular, both appeal to outcomes or results. Insofar as consequential considerations are tied to voluntary decision-making, they are moral ones. Insofar as they may be distinguished from voluntary choice, they are nonmoral.

10. There are exceptions. William Bennett's *The Book of Virtues* (1993) received great acclaim at the end of the twentieth century. In addition, numerous school systems have moved to adopt codes of virtue in light of the growing incidence of school violence.

11. A parallel here is drug abuse (Biological Sciences Curriculum Study, 2001), which in the nineteenth and early twentieth centuries was understood through a moral character model.

CHAPTER 8

THE CONTEXT OF DISEASE

One of the difficulties of writing about disease is that our understanding of disease changes with time. We tend to think of the scientific core of medicine as unchanging. Medicine, as a science, speaks about what is, not about what may be for the moment but inevitably will change. However, much has been made recently of the changing character of medicine and science. As the contributions of Ludwik Fleck (1979 [1935]) and Thomas Kuhn (1970 [1962]) to the history and philosophy of science and of medicine have shown, there are no such things as neutral, naked, and bare facts. Facts always appear interpreted within the embrace of theoretical frameworks, whether or not these frameworks are formally or informally developed as scientific accounts. In addition, they are always given in a particular socio-historical context. There is no timeless or contextless account of reality, including disease, or at least there is no such interpretation available to humans. This chapter argues that disease is *contextual*. In so doing, it explores the character of contextualism in medicine, why contextualism is not relativism, and how conflicts among competing interpretations of disease may be resolved. In the end, a *localized* account of disease is offered.

The argument here is in some sense not new. It restates what has already been argued in this project, that our understanding of disease is a function of ontological, epistemological, and axiological commitments. In other words, what we call "disease" turns on commitments regarding what it is, how we know it, and how and why we value it. Yet, the position forwarded is in some sense new. It is new in that it treats the commitments or components of disease together. Ontological commitments concern the nature of the objects of knowing while epistemological ones concern how we know the objects of knowing. Epistemological commitments concern how we know what we choose to know, while axiological ones concern how we choose to value what we know. Ontological commitments concern objects of choice and axiological ones concern choices regarding objects that are known. Thus, as this chapter argues, the commitments presuppose one another, thus creating contexts (or demarcation of reality, knowing, and valuing) to locate particular accounts of disease.

1. CONTEXTUALISM AND DISEASE

As Chapters 3 through 7 indicate, an historically and culturally unconditioned appreciation of disease is practically and theoretically impossible. In *Genesis and Development of a Scientific Fact* (1979 [1935]), Ludwik Fleck[1] argues that facts must be understood in terms of their historical and psycho-social presuppositions. He challenges the claim that "facts" are independent of the contexts in which they are

107

fashioned. He applies the Kantian insight (Grene, 1974) that the known is a product of the knower. But, unlike Kant, who portrayed the process in an ahistorically- and asocietally conditioned fashion, Fleck embeds scientific thought and scientific facts in particular historical and societal contexts. In this, he goes beyond Hegel's historical appreciation of the categories (1970 [1830])[2] and places the intersection of the knower and the known in a particular socio-cultural relationship.

Fleck elaborates his argument around a case study of syphilis. Syphilis, he argues, can be understood only if one assigns a temporal reference point indicating a particular thought collectives' understanding of the disease concept. In the course of time, the character of the concept of syphilis changes from the mystical, through the empirical and generally pathogenetical, to the mainly etiological. The transformation allows for a rich resource of fresh details and a loss of seemingly irrelevant ones. It is brought about by more specific and reliable observations of the clinical condition as well as of the change in the clinical condition brought about by the use of therapeutic interventions, i.e., mercury followed by antibacteriological agents. Theoretical and practical elements, the apriori and the purely empirical, mingle with one another, thereby resulting in a clinical concept of syphilis.

As the subtitle of his book indicates, Fleck is concerned with the genesis and development of scientific facts, which are understood in terms of a particular *Denkstil* or thought-style, i.e., "the special carrier for the historical development of any field of thought, as well as the given stock of knowledge and level of culture" (1979, p. 39). They are understood as well in terms of a particular *Denkkollectiv* or thought collective, i.e., a community of persons mutually exchanging ideas or maintaining intellectual interaction (1979, p. 39). To quote Fleck:

> Facts are never completely interdependent of each other. They occur either as more or less connected mixtures of separate signals, or as a system of knowledge obeying its own laws. As a result, every fact reacts upon many other. Every change and every discovery has an effect on a terrain that is virtually limitless. It is characteristic of advanced knowledge, matured into a coherent system, that each new fact harmoniously--though even so slightly-- changes all earlier facts. Here every discovery is actually a recreation of the whole world as construed by a thought-collective (1979, p, 102).

Put another way, medicine is a collective endeavor. Results of medical scientists and practitioners are scrutinized by the medical community. Medical findings are reported and subject to peer review in journals and at meetings. Communication must use standardized descriptions so that results are meaningful to any informed person to whom they are communicated. The division of labor in medicine provides further opportunities for the critique of medical findings and the development of multiple lines of evidence and helps to build a more complete explanation of a clinical finding. On this analysis, medicine attempts to be authoritative, but not authoritarian.

Fleck's interpretation of the history of medical science has been applied to the "facts" of the natural sciences through Thomas Kuhn's *Structure of Scientific Revolutions* (1970 [1962]).[3] Thus as Kuhn (1970 [1962]) indicates, there is no univocal sense of oxygen. If one holds that it was Priestly (1733-1804) who discovered oxygen, one must recognize that Priestly thought he had isolated dephlogistoned air (according to eighteenth century phlogiston theory of combustion). If one credits

Lavoisier (1734-1794), one must acknowledge that Lavoisier believed that he had discovered the necessary condition for acids (Kuhn, 1970, pp. 39-40). If their claims are understood narrowly, both Priestly and Lavoisier could be considered in error. Qualifications such as "they were correct in part" will be necessary each time any and all scientific or medical research develops further.[4]

These analyses of the nature of thought-styles, thought-collectives, and paradigms have led to a substantial reassessment of the contextual character of medical knowledge such as disease. Lawrie Reznek (1987) offers a contemporary justification of a contextual account of disease. For him, contextualism characterizes "a domain of discourse where the truth of the sentences varies from one system to the next, because the sentences contain relational terms such that the truth conditions also vary from system to system" (Reznek, 1987, p. 169; also see DeRose, 1992). Contextualism allows us to account for variations in disease status. As Reznek says: "We cannot decide whether a judgment about disease-status is true without considering the relation of the condition to the organism, and the relation of the organism to the environment" (Reznek, 1987, p. 169). Dyslexia is not a disease in a pre-literate environment, but in a literate one. Insensitivity to growth hormone is not a disease among pygmies, but is one among the Masai (Reznek, 1987, p. 169). What can be inferred from this is that the truth and falsity of a sentence "X is a disease" varies with the sort of organism that has the condition, and with the sort of environment in which that organism lives.

Figure 11 illustrates the relation among knowing, treating, and context (e.g., society, community, politics, law, historical period, religion, etc.).

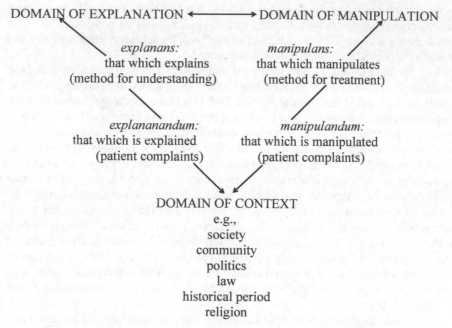

Figure 11: Relations Among Knowing, Treating, and Their Contexts

Knowing and treating interplay in the context of particular and sometimes different settings. While medicine always function in a social context, it is sometimes greatly influenced by the law (e.g., the diagnosis and treatment of communicable diseases), other times by politics (e.g., the diagnosis and treatment of AIDS), and other times by religion (e.g., clinical definitions of death). The point is to be aware of such influences and take those into consideration in one's analysis and/or actions.

        Put another way, disease is relational. To begin with, note the assumptions of rationalism and empiricism (Chapter 4). Taken together, rationalism and empiricism, assume that the universal features of knowledge acquisition by individuals are based on the presuppositions that all individuals are the same, either in capacity or structure. More specifically, rationalism assumes that individuals are all rational, and empiricism assumes that they have the same sensory or perceptual capacities. As the prior chapters have argued, such an assumption about the universal character of objects and knowledge is questionable. If knowledge is universal, it is asocial, or independent of particular communal structures. But, according to Wartofsky (1992, pp. 133-134), such a position contradicts the very position affirmed by the rationalist and empiricist. Rationalists and empiricists affirm a sociality of knowledge insofar as *all* human beings are considered to share in these universal features of reason or sense perception.

The same can be said of metaphysical and axiological accounts of knowledge and action. Metaphysical realism and metaphysical anti-realism both assume that all of reality shares in the universal features of what reality is. Neutralism and normativism, and objectivism and subjectivism, assume that all human beings share in the universal features of how humans assign significance and worth.

Our challenge, then, is to make explicit the assumptions implicit in medical epistemology, ontology, and axiology. In doing so, we recognize that the relational character of medical knowledge of disease is a function of several givens, which need to be made explicit (Wartofsky, 1992, p. 133). First, medical knowledge as human knowledge assumes biological *a priori* conditions, e.g., a certain development of brain and neural system, or the organs for speech and other forms of symbolic communication that are shared by humans, thus making communication possible. Second, medical knowledge as institutional knowledge assumes collective *a priori* conditions, e.g., some pre-existing form of social life, a linguistic community, the growth of techniques of production and of social organization. Third, medical knowledge as historical knowledge assumes concrete forms of human interaction that are typical of a given society in a given period that is influenced by former periods. Yet, according to Wartofsky (1992, p. 133), these three conditions give only the conditions of the possibility of knowledge in medicine. It does not specify *how* medical knowledge is generated, or *why* it is generated; nor does it give us, as yet, any concrete or simple understanding of the knowledge itself. Therefore, every particular form of knowledge has to be specified. It has to be specified with regard to the needs that the knowledge serves and the recognition of those needs that is expressed in the conscious purposes to which this cognitive acquisition is addressed.

As a consequence, one can say that medical knowledge is teleological. It is acquired and used for the sake of some end. But this end is not of one kind; rather it has multiple expressions. One can say that medical knowledge is the discovery of scientific truth or human good, the pursuit of a pleasurable or financially lucrative life, and/or a key to understanding human freedom or determinism. Each may be seen as a conscious recognition of particular human needs, which are shared by others.

Thus, what distinguishes medical knowledge from biological, historical, or moral knowledge are the distinctive ends medicine serves. Consider that members of the health care professions are dedicated to the following sorts of goals, not all of which are always in harmony: they (1) serve the health care needs and desires of individuals, (2) support the health care needs and desires of societies, (3) engage in their profession to gain income and prestige, (4) aid the profession in being self-perpetuating, and (5) aim at the acquisition of knowledge (Engelhardt, 1996, pp. 292-293). In addition, in this era of managed care, they also (1) compete in the market, (2) function as middle-level managers in health care situated between patients and upper-level managers, and (3) serve as translators of clinical outcome assessment data for patients, health care managers, and insurers (Cutter, 2002a).

Note some of the tensions that may arise. The good of the individual and institution may compete (as might occur when a clinician is rewarded for conserving resources in managed care). The good of the individual and society may be at odds (as in the case of mandatory disclosure requirements). The pursuit of individual gain (such as financial rewards) may lead to undermining the status of the profession. The pursuit of knowledge (in the case of research) may conflict with the interests of the individuals being treated. The point is that medical knowledge and the goals it sets out to achieve are varied and complex (Callahan, 1999; Peppin, 1999; Bondeson and Jones, 2002).

Our task, then, is to examine what the norms are of the practice of medicine itself. Consider that today health care is practiced by teams of diverse disciplines and not individual isolated practitioners. It is institutionally based, and the "science" now includes health services research, quality review mechanisms, complex networks of informational systems, and logical strategies for the coordinated use of expensive, institutionally based technologies. The science is different from the biomedical sciences and involves multiple levels of management, organization, specialities, and professions (Agich, 1999). Further, health care no longer involves an ongoing patient-physician relationship (Peppin, 1999). American patients and physicians are extremely mobile. Because health insurance in America is typically purchased in the workplace, and workers typically change jobs over their lifespans, American workers/patients experience numerous changes in their health care delivery. In addition, given American consumerism, American patients regularly shop for their providers and change providers when dissatisfaction occurs. The point is that medicine is changing, and with this change comes new challenges. The task, then, is to see what the profession and the practice of medicine specifies as knowledge and the ends it seeks to achieve.

What one will find is that one will have to rethink the meaning of disease. Disease is not simply a pathology either of an organ or of an organism, but of the larger context in which the organism grows, survives, or dies. The mediating link between the biological organism as such and the social, historical, and cultural contexts of the organism is the social or institutional forms of medical practice and other contexts of individual and collective practice. Institutional forms of medical practice include the hospital, clinic, health maintenance organization, medical research center, and particular specialities and sub-specialities of medicine, e.g., pubic health, family medicine, genetic medicine). Other contexts include the life world of the practitioner, his or her social class, the way he or she earns a living, and the prevailing value-structure or structures, as well as the relations between medical research and applied practice and between basic research in science and medicine. In the mediation between the organism and its contexts, to borrow from Wartofsky, "the organ and the organism...have become transformed into social entities, and may no longer be considered abstractly at the biological level alone..." (1975, p. 70). Disease is, in short, contextual.

2. WHY CONTEXTUALISM IS NOT RELATIVISM

It may worry some that relativism appears to be compatible with the position advanced by this work. Such a conclusion is in err, as the following discussion details.

To begin with, relativism is the belief that knowledge is determined by specific qualities of the observer, including age, race, gender, and cultural conditioning. For example, the Sophists (e.g., Protagoras [481-411 B.C.], Curd, 1996) claim that place of birth, family habits, personal abilities and preferences, religious training, age and so forth control an individual's belief, values, and preferences. Based on this view, we need only accept what seems true at the moment, according to our culture. More extreme Sophists claim that even within the same culture, individuals have their own beliefs.

Two basic variants of relativism are distinguishable: cultural and individual. Cultural relativism is the belief that all values are determined by the ideas, concepts, and beliefs humans in community live by. Humans in community tend to assume similar perspectives, and likes and dislikes, in order to live harmoniously together. Monthly magazines that feature home, dress, art, and music fashions illustrate trends well. Humans from different communities often diverge in perspective, likes and dislikes, in order to distinguish themselves from each other, to give themselves an identity, and to respond appropriately to their environment.

Individual relativism carries the logic of cultural relativism to a more refined conclusion. Even in the same place and time, good and bad, and right and wrong, are relative to the unique experiences and preferences of the individual. There is no unbiased way to say that one standard is better than another for the standard used to make a given claim is in itself the reflection of a preference, *ad infinitum*. No matter how far back we push "ultimate" reasons, they always reduce to individual preferences, likes or dislikes.

Relativism appears to be an attractive position in part because relativism seems to account for a culture's diversity and commitment to tolerance. Unfortunately, these advantages cannot be achieved because relativism fails on two counts, namely, conceptual and practical. Conceptually speaking, relativism advances the view that "truth is relative." If the claim is taken seriously, then it backfires on itself: if it's true, it must be false. If truth is relative, then the claim itself is relative, thereby undermining its ability to establish anything. Correspondingly, we can be certain of nothing. Such a claim is self-refuting. Relativism commits one to the thesis that one and the same proposition can be true in one system and false in another. This is equivalent to endorsing and condemning one and the same state of affairs. Practically speaking, relativism fails to take into account what common sense tells us: that we do act as if there is understanding or certainty in our world. We rely on math and science to build our world and are often quite successful in doing so. We do think and act as if things matter. We vote, crusade for causes, make sacrifices for various ideals, and hold each other responsible for personal actions. Practically speaking, then, there is really no such thing as a consistent relativist.

Disease is not, then, relative. Disease describes fundamental processes and causal relationships that are grounded in natural phenomena. As such, disease has predictive power. Because disease is constructed according to rigorous criteria, it should, for the most part, withstand the fad and popular beliefs of any particular society at any point in history. It does so because medical explanations, like scientific ones, require evidence collected according to a rigorous set of rules (American Association for the Advancement of Science, 1989; National Academy of Sciences, 1998; Chapter 4). Disease is built through intellectually disciplined consensus, but not through the consensus of voting (Moore, 1992; McInerny and Moore, 1993)[5]. Yet the path by which medical discovery proceeds is conditioned by history and culture. Social and political values do help to shape disease, particularly the agenda of research at any given time, because modern medical research often requires large amounts of money and is therefore open to public scrutiny.

3. A CENTRAL ROLE FOR NEGOTIATION

As this investigation illustrates, the ways of knowing and treating disease have different expressions and different values depending on the stakeholders. In medicine, we come together to speak to these expressions and values. Systematic programs for investigating disease and assessing treatment options are in principle not the undertakings of isolated individuals. They are the result of communities of physicians, patients, and the general public coming together to negotiate how clinical reality can and should be interpreted and undertaken.

The negotiation metaphor offers a schema for assessing various claims regarding the ways in which we understand and manipulate disease. Negotiation (L. *negotiatio*, from *negotiari*, to carry on business) as a procedural expression of respect is often the only way to resolve disputes where there is disagreement about what is proper to do (Engelhardt, 1996). In the law, "negotiate" means to conduct communication or dialogue with a view toward reaching a settlement or agreement. It is the *process* of arranging the terms and conditions of a transaction (e.g., contract) (Black, 1979, p. 934).

Negotiation in medicine takes on either a formal and informal character. Formal discussions regarding the nature of disease are found in the literature and in special organized forums such as conferences and committees. Recall, for example, the work of the International Committee on Viral Taxonomy (Varmus, 1989) and its work on naming the AIDS virus. This kind of formal negotiation in science and medicine is seen as well on committees that create and/or revise clinical classifications in speciality clinical groups. Discussions regarding what constitutes disease take on as well an informal character and are found in the offices and laboratories of clinicians and scientists. Discussions concerning what to focus on, how to name, whether to fund, and so on typically begin on an informal note and often go hand-in-hand with formal discussions.

This is not a plea for understanding and classifying diseases by referendum or town meeting, as may be the case of politics (McInerney and Moore, 1993; Cutter, 1992, 1998). This is instead the claim that choices among different understandings of reality within medicine are matters of communal agreement and interest. This recognition underscores our choices and indicates our responsibilities as individuals who not only study clinical reality in order to know it (Chapters 3 and 4), but know it in order to manipulate it, and manipulate it in order to know it (Chapter 5). As a consequence, communities of clinicians, scientists, patients, and related parties negotiate the characterization of clinical reality when the interests of participants do not coincide.

Notions of correct explanation depend on accepted rules of evidence and inference within particular communities.   Notions of good explanation depend on functional, aesthetic, or instrumental analysis.   Functional values tell us what ideals of activity are proper to an organism. Aesthetic values tell us what ideals of form and grace are worthy of achievement. Instrumental values tell us how to get from means to end, where the end of the maximization of benefit and minimization of harm, cost, or burden.   Notions of right explanation depend on moral analysis, and can either specify the right independently of that which is good to do, or interpret that which is right to do as that which maximizes or achieves the good.   Agreement may be widespread. One might think here of a long-standing patient-physician relationship in which both parties (1) understand the nature of the patient's problem, (2) wish to achieve a specific therapeutic good (in a functional, aesthetic, and instrumental sense), and (3) respect each other (in an ethical sense). In contrast, disagreement may abound. One might think here of consensus groups in which experts from diverse clinical specialities are brought together to decide on the diagnosis and course of treatment for a clinical condition such as AIDS (Chapter 2).   (1) There is not a univocal understanding of what constitutes the problem, (2) there are conflicts regarding what goods are at stake (functional, instrumental, and aesthetic), and (3) there are disagreements between respecting persons (by maintaining confidentiality of medical records) and serving the public good (by disclosing information for the public good).

Given these constraints, negotiation becomes the only credible approach to resolving a wide range of disputes.   In order for nature to serve as a criterion for resolving disputes with ethical and political overlay, it must be shown to be morally normative.   One must be able to show that the general tendencies of nature were established by God to guide humans to their proper goals. To establish such a proposition would require special metaphysical premises not open to general secular defense. This is a problem in secular communities committed to peaceable resolution of controversy. Reliance upon negotiation thus becomes a reasonable goal for all those wishing to resolve controversies when consensus is not available. When matters are put within this more abstract, procedurally-oriented context, there is the possibility for negotiating among partisans of different disease understandings and recognizing that values play a central role in such negotiations.

This is not to suggest that trans-cultural accounts of disease are impossible. Clinical communities can and do agree on a great many explanatory accounts because they are bound by the constraints of scientific methodology and influenced by similar notions of what makes human biological and psychological life pleasant versus unpleasant, good versus bad, and right versus wrong. Nevertheless, there will be disagreements.   A result of this sobering assessment of our limitations is that differences may often be resolved only by negotiation. This is unlikely to be a pleasant conclusion, but taking human limitations and human freedom seriously is always challenging.

4. DISEASE, UNCERTAINTY, AND MYSTERY

This study offers a procedural analysis, a schema for assessing various claims regarding the ways in which we explain and manipulate disease without endorsing any particular vision. These claims are, as we have seen, metaphysical, epistemological, and axiological. They correspond to three major features of disease concepts, which relate to each other in various and complex ways. One can see these features as providing the bases for a typology of diverse accounts of disease such as reviewed in Chapter 1. But they only provide the bases. Even after the success of this account, one will still face a choice among particular accounts of disease. One will have to learn the rules of evidence and inference accepted by those working in the particular community one wishes to join. One will have to reflect on how to weigh certain judgments vis-a-vis particular clinical outcomes, which involves listening to and understanding those involved in select clinical discourses. One will, in addition, have to accept that disease concepts evolve and change with new knowledge and recognitions of previously unrecognized valued and dis-valued states of affairs. Given the futility of discovering clinical reality and a single sense of disease, we begin to appreciate that there will at times be differing, if not competing, interpretations of disease.

It is no wonder, then, that uncertainty, and often mystery, characterize disease. Kendell (1975, p. 10) claims that the concept of disease originated as an explanation for the onset of suffering and incapacity in the absence of obvious injury. As Reznek explains, conditions without obvious causes cried out for a different sort of explanation compared to injuries like fractures and burns. The explanation for the latter occurrences was located in the external cause, but in the other conditions where harm resulted, some intervening variables had to be postulated. In short, accounts of disease emerge out of quests to know and to treat human suffering, the basis of which was uncertain.

One might note that there are different senses of uncertainty that arise in the process of framing disease. There is the sense of uncertainty that refers to a lack of knowledge. While there was strong evidence of the infectious nature of AIDS, no viral agent for AIDS had been identified in the early 1980s. Even with the identification of the viral agent in the mid-1980s, clinico-researchers still did not understand how the agent attacked the host and lead to immunological breakdown until the 1990s. Even then, as new discoveries answer some questions, they inevitably raise others, such as why the disease process differs among patients. In addition, there is a sense of uncertainty that is often involved in reaching a diagnosis. In the early 1980s, the diagnosis of what came to be known as AIDS was made not on the basis of an etiological laboratory test, but instead on the basis of signs and symptoms. As a result, it was often more difficult to make a diagnosis quickly and well before onset of symptoms. But even with etiological laboratory tests, questions are raised regarding false-positives and false-negatives. Then there is a sense of uncertainty that arises in probabilistic thinking, which is common in medicine. For many clinical conditions (particularly genetic ones and even cases of HIV exposure that do not lead to AIDS), there are empiric risk figures that estimate the increased risk of a disease. Such figures are usually not 100% and thus leave the impression that the clinical information is uncertain. Beyond this, risk estimates concern the distribution of traits in populations, and applying this information to individuals is a matter of judgment, not algorithmic reasoning. Finally, there is a sense of uncertainty for some when values are involved, for values are not universally shared by individuals. For instance, the American Psychiatric Association (1980, p. 281)[6] agrees that homosexuality is not in and of itself a disease, but may be considered a psychiatric disorder if and only of the patient sees it to be a problem (i.e., disvalues it). In short, uncertainty characterizes medical knowledge and expresses itself in various ways.

For many, the complex and sometimes uncertain character of disease is not easy to accept, especially when one is in the business of making concepts clear (as in the case of philosophers, scientists, and clinicians) or the position of patient or advocate of a patient wanting answers. Part of the difficulty we face is that we have inherited traditions of thought (e.g., the Enlightenment) that promise answers and clear ones at that (Sassower, 1993). Yet, the medico-scientific tradition, one heavily built on inductive reasoning, is unable to deliver True knowledge. One of the main points forwarded by this study is that we must begin to accept that disease, not unlike other aspects of the human condition, is a label fraught with ambiguities and mystery.[7] It is the intent of this work to come to terms with ways to think about disease, absent expectations that disease can be known fully.

5. LOCALIZING DISEASE

So far, the analysis argues that, procedurally speaking, disease refers to descriptions of pathological psychosomatic processes that occur in the human species and bring patients to the attention of health care professionals. In this way, disease involves three components: descriptive, prescriptive, and social. In terms of description (Albert 1988, et al.), disease results from careful observation and well-reasoned arguments involving the formation and testing of hypotheses (or proposed explanations). The clinician starts with observations of the disease, moves to propose an explanation, tests the explanation against evidence from the presenting disease, and, with each presentation of the disease, tests new evidence against existing explanations. Through this method, medicine has demonstrated remarkable power to describe clinical problems in terms of symptoms, underlying factors, and relations that are open to a range of interpretations.

In this process of explanation, disease labels or classifications give rise to the development of norms. Such norms function prescriptively. They serve as the basis for judgments about how individuals with certain conditions ought (not) to be, (not) to act, and so on. Individuals ought to be able to function according to ideals of activity that are proper to an organism and those of form and grace that are worthy of achievement. Furthermore, we decide how to act, what to strive for, and what to resist in light of such norms. We ought, for example, to provide treatment to those who are suffering. In this way, disease functions as a treatment warrant for those qualified to offer a response. Disease-norms are clusters of characteristics and abilities that function as standards by which individuals and their conditions are judged to be "good" and "bad" instances of particular human conditions.

As framed here, the descriptive level of analysis in discussions of disease is tied to a prescriptive level. Facts and values, as Engelhardt (1996, Ch. 5) and others (e.g., Fleck, 1979 [1935]) teach us, interweave in complex ways. Observations in medicine are always ordered around theoretical commitments, including judgements concerning how to select and organize evidence into explanations. Further, observations are always ordered around evaluative commitments, including those concerning what phenomena are assigned significance in terms of functional, instrumental, and aesthetic values and what actions are appropriate in order to achieve certain goals (e.g., minimizing suffering, respecting patient autonomy).

Moreover, the prescriptive force is backed by social sanctions, fashioned in light of what goals are seen to be worthy of achievement in our collective lives. If one conforms to a disease-norm, one increases the probability of being a recipient of social support (Parsons, 1951). If one diverges from such norms, one lessens the chance of being a recipient of social support. In short, disease involves descriptive, prescriptive, and social commitments.

Yet, disease involves more. Even after the success of this account of disease, one still faces a host a choices regarding particular accounts of disease. One will have to learn the rules of evidence and inference accepted by those working in a particular community one wishes to join. One will have to reflect on how to weigh certain judgements vis-à-vis particular outcomes, which involves listening to and understanding those involved in select clinical discourses. One will, in addition, have to accept that our understanding of disease evolves and changes with new knowledge and recognition of previously unrecognized values and disvalued states of affairs. Given the futility of discovering clinical reality and a simple sense of disease, we begin to appreciate that there will at times be differing, if not competing, interpretations of disease.[8]

On this account, disease is what I will call a "localized" concept. To begin with, disease is localized in terms of what and how it describes. AIDS, for example, can be correlated with genetic, immunological, anatomic, and social variables, depending on whether one is a molecular biologist, immunologist, oncologist, or public health official (Chapters 2 and 3). The construal will depend upon the particular clinician's appraisal of which etiological variables are most amenable to the framework the observer brings to bear on the observation. For example, an immunologist may decide that the major factors in AIDS is the T-cell count. The public health official may decide that the basic variables in AIDS are elements of a lifestyle that include unprotected sex or use of unclear needles. Much depends on how the facts are nested.

In addition, disease is localized in terms of what and how it values (Chapters 4-7). Disease exists in order to cure, to care, and to intervene. Since we understand disease in terms of responding to patient complaints, it follows that all disease takes place against a background of value presuppositions. Simply put, functional values tell us what ideals of activity (e.g., functioning immune system) are proper to an organism. Aesthetic values tell us what ideals of form and grace (e.g., a body without infection) are worthy of achievement. Instrumental values tell us how to get from means to ends (e.g., strong immune system, which results in health), where the end is the maximization of benefit and the minimization of harm. Since all may not share in the same view regarding what ideals are proper to be achieved, ethical values are at stake as well. To diagnose, predict, and treat, while taking proper regard of patients, requires acknowledging the range and specific expressions of values embodied in how we understand disease.

Finally, disease is localized in terms of cultural forces (Chapter 8). We frame disease not only in terms of the relation of the condition to the organism, but in terms of the relation of the organism to the environment (Reznek, 1987, p. 69). AIDS is considered by some not only a medical condition but a spiritual one (Chapter 2). Dyslexia is not a disease in a pre-literate environment, but in a literate one. Anorexia nervosa is a diagnosis in a resource-rich culture that places positive value on thin body image, but not one in a culture where starvation is common. Even within certain clinical categories, such as depression, seemingly universal afflictions can differ profoundly from culture to culture (Osborne, 2001, p. 100) such that the developers of the Diagnostic and Statistical Manual of Behavioral Disorders IV (American Psychiatric Association, 1994) introduced a new category of psychiatric illnesses known as "culture-bound syndromes." In short, the descriptions, values, and socio-political influences that enter into how we understand and manipulate disease are localized and depend on the perspectives taken by those providing the account.

It does not follow from this analysis of the localized nature of disease that "anything goes." Although we may lack a singular interpretation of disease, relativism is not the only option. There are constraints or limits on interpretations of disease. There are similarities or common grounds. The constraints on interpretation in clinical medicine are the reality of patient suffering or pain (Chapter 3), the methods we use to know suffering (Chapter 4), the responses we render (Chapter 5), the acts of evaluating suffering or pain (Chapters 6 and 7), and the influences of socio-political forces (Chapter 8). These and related constraints are entailed in the account of naturalism that frames this analysis. On this view, the world is part of a natural order that humans did not create and that can be studied using the methods of justification and explanations supported by the sciences.[9] Despite the constraints, though, there is much that can be said about disease.

6. CLOSING

Our understanding of disease is located in diverse ontological, epistemological, and axiological commitments, which are expressed in particular contexts. The chapters that follow apply the analysis thus far presented and consider clinical problems that are commanding greater attention at the turn of the new millennium, namely, genetic disease and women's disease.

CHAPTER ENDNOTES

1. Note that Fleck does not refer to himself as a contextualist.

2. There are many roots in the history of thought to which one could appeal to show the origins of the realization of the contextual character of facts. Immanuel Kant (1724-1804) grounds the character of the known in the characteristic of the knower:

> Now all experience does indeed contain, in addition to the intuition of the senses through which something is given, a *concept* of an object as being thereby given, that is to say, as appearing. Concepts of objects in general thus underlie all empirical knowledge as its *a priori* conditions. The objective validity of the categories as *a priori* concepts rests, therefore, on the fact that, so far as the form of thought is concerned, through them alone does experience become possible. They relate of necessity and *a priori* to objects of experience, for the reason that only by means of them can any object whatsoever of experience be thought. (Kant, 1929 [1789] B126, p. 126)

Facts are the products of the knower, not just the known. To use the Kantian idiom, we do not know things as they are in themselves, as they are totally apart from our categories, but only as they are given to us through our categories of understanding. "What the things-in-themselves may be I do not know, nor do I need to know, since a thing can never come before me except in appearance" (Kant, 1929 [1789], A277-B333, p. 286)

Following Kant, G.W.F. Hegel (1770-1831) expands on the Kantian view that reality is created, not simply discovered. He recognizes that the relation between the knower and the known evolves through time. Categories should no longer be seen as atemporal but appreciated within an historical development.

All cultural change reduces to a difference in categories. All revolutions, whether in the sciences or world history, occur merely because spirit has changed its categories in order to understand and examine what belongs to it, in order to possess and grasp itself in a truer, deeper, more intimate and unified manner (Hegel, 1970 [1830]). Hegel also appreciates that the development of knowledge is not simply linear, but marked by radical changes or revolutions.

3. Kuhn's term "paradigm," unlike Fleck's "thought-style" and "thought-collective," succeeds in bringing together a set of developing insights regarding the contextual character of knowledge (Hanson, 1958; Gutting, 1980).

4. King (1982, pp. 227-244) makes a similar point when he defends the legitimacy of bleeding in the history of medicine. Numerous commentators dismiss bleeding as a crazy attempt on medicine's part to approach disease and illness. King responds that bleeding has played an important role in medicine. One has to take care in considering the contexts and goals of the scientists, health care providers, and patients. For a contemporary scientific defense of bleeding, see Gardell et al. (1990, 1991).

5. Also see Chapter 4, "On Clinical Method."

6. Homosexuality has develop from an instance of socio-pathic personality disturbance in the first *Diagnostic and Statistical Manual of the American Psychiatric Association* (DSM-I) (American Psychiatric Association, 1952, pp. 38-39) to a "personality disorder" in DSM-II (American Psychiatric Association, 1968, p. 44), and to an instance of pschosocial dysfunction in DSM-III (American Psychiatric Association, 1980, p. 281) under the taxon "ego-dystonic homosexuality," and then to an instance in DSM III-R (American Psychiatric

Association, 1987, p. 296) of sexual dysfunction under the obscure taxon "sexual disorder not otherwise specified." In DSM-IV (American Psychiatric Association, 1994, p. 538) and DSM-IV-TR (American Psychiatric Association, 2000, p. 582), homosexuality remains under the same rubric as in DSM-III-R.

7. Here, I think, is an important contact point between science or medicine and theology. For those in faith traditions, mystery breeds reverence, in this case the revering of the natural order which is far more complex than we presently understand. For those in faith traditions that provide a key role for reason, mystery in science and medicine provides an opportunity to come to know more fully the natural order.

8. Aronowitz (1998) makes a similar point. He argues that each disease has a particular identity. This in includes medical aspects about cause, signs, symptoms, prognosis, and treatment, but also the socio-historical factors that bring about its development. Tuberculosis, for example, cannot be understood apart from its social and historical import, which varied from focus on manifesting signs and symptoms (e.g., a distinctive cough) to having microscopic tubers in the body.

9. Granted, the ontology and epistemology needs much further development but will not be pursued here.

CHAPTER 9

CONCEPTS OF GENETIC DISEASE[1]

The Human Genome Project (HGP) has spawned a tremendous explosion in research in genetic science and medicine. A major impact of the HGP will be an evolution in the way we think about disease and normal physiology. What follows is an analysis of the concept of genetic disease, in light of the analysis that has taken place in this work, particularly in Chapters 3, 4, and 5. In this way, the analysis in this chapter carries out the lessons of this work by considering the nature of disease--in this case genetic disease--and how we know it, and how it reflects what and how we value. It illustrates how this work has application in current debates.

1. WHAT WE KNOW

1.1. The Human Genome Project

The Human Genome Project (HGP) is a federally-funded three billion dollar, fifteen-year enterprise that began in October 1990. The HGP has two goals: (1) to develop detailed maps of the human genome and the genome of several other well-studied organisms (e.g., bacteria, yeast, fruit fly, and mouse), and (2) to determine the complete nucleotide sequences of these genomes.

A geographical analogy may be helpful for understanding how gene mapping and sequencing will be done (Walters, 1997, pp. 223-224). If you wanted to make a physical map of the United States, you might first use a satellite photograph of the whole country. This satellite photo corresponds to locating the 23 pairs of human chromosomes. You might divide the country into 1,000 regions, each with a certain number of square miles. For this more detailed map, you might take photographs taken from airplanes. Similarly, scientists have begun to divide the human genome into major regions, each of which consists of perhaps 30,000-35,000 units or base pairs (i.e., a purine [i.e., adenine or guanine] or a pyrimidine [i.e., thymidine or cytosine] on one strand of the DNA that forms hydrogen bonds with a complementary base on the other strand). In our geography project, there might be certain regions that require special attention even at this stage of mapping, for example, major metropolitan areas or areas with potentially dangerous geographical faults. In a parallel way, scientists have discovered that certain health- and disease-related genes are located in particular regions of particular chromosomes. They have begun to investigate these regions in greater detail at this stage of mapping. The final stage in the geography project might be a highly detailed map that indicates streets or even individual buildings on those streets. These fine details correspond to knowing the sequence in which some or all of the 3 or so base pairs are arranged in human chromosomes.

One essential part of the HGP is the development of technology that will enable more efficient and effective studies of the genome. A second important piece is the development of systems for electronic database storage and management to handle information the HGP will generate. Such information is tremendous. A written record of the human genome sequence is estimated to require approximately 200 telephone books of 1,000 pages each (Biological Sciences Curriculum Study, 1996, 2003). In short, one of the major goals of the HGP is to map and sequence the human genome.

This leads us to ask what is a human genome? Human beings usually have 23 pairs of chromosomes in the nuclei in every cell. In turn, the chromosomes contain genes that code for proteins and other products (e.g., polypeptides) that our bodies need to function, as well as intervening stretches of DNA whose function is not yet understood (and has been referred to as "junk DNA"). The smallest coding units within the genes and intervening sequences are called codons. It is estimated that our 23 pairs of chromosomes contain approximately 30,000 genes and 3 or so billion base pairs.[2] Scientists seek to learn more about the structure and function of human genes, chromosomes, and extra-nuclear DNA, which taken together is called the *human genome*.

For purposes of study, the HGP defines the human genome as a single haploid (i.e., single) set of nuclear chromosomes, plus the mitochondrial genome. Here the mitochondrial genome constitutes the genetic material found in mitochondria, the organelles of cells of eukaryotes (i.e., organisms whose cells contain a limiting membrane around the nuclear material), which are important in energy-related reactions. Mitochondrial genetic material can replicate independently of the genomic DNA found in the nucleus, and cells have many mitochondria. In humans, mitochondria are transmitted maternally, a phenomenon that some refer to as nontraditional or Non-Mendelian modes of inheritance (Biological Sciences Curriculum Study, 1997). In short, the HGP seeks to understand the human genome, which involves a search to explain the structure and function of genes.

1.2. Genetic Testing

The first application of new knowledge in genetics appears to be diagnostic. Tests are being developed for large numbers of diseases[3], and these tests will be used in genetic testing and screening. Here, genetic testing refers to the use of diagnostic procedures or determinations of the presence or absence of one or more genetic traits or conditions of an individual. More specifically, genetic tests mean the analysis of human DNA, RNA, chromosomes, proteins, or other gene products to direct disease-related genotypes, mutations, phenotypes, or karyotypes for clinical purposes (National Institutes of Health/Department of Energy, 1992). Genetic screening refers to the use of genetic tests in programs that are intended to reach a large number of persons who have a particular genetic trait or condition. Thus, genetic screening always involves genetic testing, but genetic testing does not always involve a systematic, organized genetic screening program. The discussion that follows focuses on genetic testing.

As currently practiced, genetic testing usually occurs at one of four stages of life: (1) in the neonatal period; (2) prenatally; (3) when couples are considering reproduction; or (4) when a person, on the basis of family history, recognizes that he or she has a higher-than-average risk of developing a genetic problem. Consider briefly each type of genetic test. Neonatal testing is confined to so-called single-gene disorders (e.g., phenylketonuria, hypothyroidism, sickle-cell anemia, galactosemia, homocystinuria, and maple-syrup urine disease) amenable to treatment and for which a reliable test exists. Prenatal testing is targeted toward single-gene disorders (e.g., Duchenne muscular dystrophy, cystic fibrosis, hemophilia, fragile-X syndrome, Tays-Sachs disease, and Down syndrome) and are made possible by recent advances in molecular genetics. In carrier testing (of, e.g., sickle-cell anemia, Tays Sachs, thalassemia, cystic fibrosis), the relative risk of transmitting a genetic defect is assessed. For example, if only one member of the couple carries the gene for a recessive genetic disease, the couple is not at risk of producing a child afflicted with the disease but has a one-in-two chance of producing a carrier. However, if both members of a couple carry the gene for a recessive genetic disease that is not sex-linked, the couple has, they say, a 25 percent chance of producing an unaffected child, a 50 percent chance of producing a carrier, and a 25 percent chance of producing a child who will have the disease. Predictive or presymptomatic testing allows a person with a family history of a certain disorder (e.g., Huntington disease, hemochromatosis, polycystic kidney disease, early-onset Alzheimer disease, familial hypercholesteremia) to learn in advance whether he or she is at high-risk to develop the disorder later in life.

1.3. Taxonomy of Genetic Prediction

Genetic testing provides data that in turn is used to predict the onset of a clinical disorder. Managing the access and use of genetic risk information will involve addressing a range of challenges including (1) how to define the indications for and limits of predictive genetic testing, (2) how to convey prognostic information to individuals and families, (3) how to assist individuals and families in identifying, understanding, and controlling inherited health risks, and (4) how to weigh the interests of the individuals who learn their genetic health risks against the interests of society or other parties outside the therapeutic relationship who could be affected by the information (e.g., insurers, employers, educators) (Juengst, 1995).

The challenges noted above derive from a central one in genetic medicine, namely, the management of uncertainty (McInerney, 2002; National Coalition, 2002). To begin with, there is a sense of uncertainty that comes from dealing with incomplete scientific information. While there is strong evidence of heritability in many diseases, no susceptibility factors for many diseases have been identified. Tied to this, there is a sense of uncertainty that comes with challenges in reaching a diagnosis. Today, many diagnoses of genetic conditions are not made on the basis of laboratory tests, but instead on the basis of symptoms. As a result, it is often difficult to make a diagnosis quickly and diagnoses made in the past (e.g., by family members or a health care provider) may not always be reliable. Further, there is a sense of uncertainty that occurs in dealing with probabilistic risk assessments. Even when susceptibility genes are identified, the presence of those genes will not always mean that the individual will experience the disorder. Conversely, those who do not have those genes will not necessarily be free of the disorder. Finally, there is a sense of uncertainty when applying population-based data to individuals. Risk estimates concern the distribution of traits in a population. Applying this information to individuals is a matter of judgment, not algorithmic reasoning.

As the Human Genome Project progresses and linkage marker testing gives way to direct clinical testing based on DNA sequence information about the mutations in question, the management of uncertainty will change. Increasingly, as tests for genetic factors in complex diseases become available, the question will become less whether one has the mutation in question and more what that signifies for one's future. To be able to use the results of genetic research intelligently, health care professionals, patients, and research subjects will need to be able to think in terms of shifting ranges of probabilities of the influences by *both* genes and environmental factors. This way of thinking is surely against the grain of current culture's tendency to understand genetics risk factors in a deterministic[4], if not fatalistic[5], fashion (Holtzman, 1989; Childs, 1994, 1999).

To help us come to terms with different senses of uncertainty in genetic prediction, Juengst (1995a, pp. 32-3; Juengst, 1995b) provides a helpful taxonomy of six categories or senses of genetic prediction.[6] This taxonomy is offered as a way to establish the clinical significance of different categories of genetic risk assessment tests. In order of high to low degree of statistical certainty, these senses are:

1. diagnostic genetic test: confirms the diagnosis of an active disease process (e.g., mutation analysis for fragile X in developmentally-delayed children) (Roussea et al., 1991).

2. prognostic (presymptomatic) genetic test: forecasts "the emergence of a clinical health problem with a high degree of certainty" (Juengst, 1995a, p. 32) (e.g., mutation analysis for Huntington's disease in its earliest stage) (Brandt et al., 1989).

3. predictive genetic test: detects a "true genetic predisposition to a clinical health problem" (Juengst, 1995a, p. 32) (e.g., screening for PKU) (Holtzman, 1970). "[U]nless the predisposition is controlled, the clinical problem will result" (Juengst, 1995, p. 32).

4. prophylactic[7] genetic testing: detects the presence of a "genetic susceptibility" (Juengst, 1995a, p. 33), i.e., a vulnerability to a particular environmental stimulus) (e.g., alpha-antitripsin deficiency) (Stokinger et al., 1973). In the absence of the environmental stimulus (e.g., tobacco smoke), the deficiency does no harm. In its presence, the deficiency is a serious liability.

5. probabilistic genetic testing: involves a "less determined category of genetic risk assessment" (Juengst, 1995a, p. 33) (e.g., p53 mutations in Li-Fraumeni

family members) (Li et al, 1992). No strong claim "about specific course of the future" (Juengst, 1995a, p. 33) can be made.

      6.   genetic profiling: involves a category of tests that identify a "loose empirical association between a particular mutation and an increased incidence of a given health problem" (Juengst, 1995a, p. 33) (e.g., ACE and risk of myocardial infarction) (Cambien et al, 1992).

      Juengst's taxonomy of genetic testing findings provides a means to distinguish among different kinds of clinical predictions or forecasts in terms of statistical certainty. It is offered, along with others (e.g., National Coalition, 2002), as a way to combat simplistic interpretations of genetic causation and the predictive power of genetic tests.

2. TOWARD AN UNDERSTANDING OF GENETIC DISEASE

At first blush, the genetic predictive taxonomy as presented above appears to address the challenges of managing uncertainty of genetic test information. It represents a common way in which genetic counselors think about different degrees of certainty in genetic prediction. Then again, it may not address our challenges. Juengst's taxonomy of genetic prediction highlights three significant conceptual limits of genetic prediction that are worthy of consideration if we are going to understand genetic disease appropriately.

2.1. Three Concerns

First, Juengst's taxonomy of genetic prediction forges a distinction between genetic and non-genetic tests that may not be defensible. Three major reasons are often given for the distinction between genetic and non-genetic tests (National Institutes of Health/Department of Energy, 1992). (1) Genetic tests give more precise information about the likelihood that an individual will develop a particular disease. (2) Genetic tests provide information about the relatives of the individual who is tested, and conversely, the disease status of relatives provides information about the future health of the individual. And (3), genetic information about an individual is regarded as indicative of fundamental and immutable characteristics of that individual (see also Fox-Keller, 1991).

      There are problems with this distinction between genetic and non-genetic tests. (1) Genetic tests are not necessarily more predictive than nongenetic ones. Mutations are not always fully penetrant[8] in the population. The presence of the altered genes associated with Huntington disease and cystic fibrosis, for examples, exhibits less than 100% penetrance. (2) Genetic tests are not the only means, or even the primary means, of making use of the medical history of relatives in predicting the health of an individual. (3) As the first point indicates, the perception of genetic testing information as significant is based on a mistaken idea that the genetic contribution of a disease can be cleanly separated from the environmental contribution. A genetic essentialist view is untenable given much of the research in human genetics that reveals complex interactions among genes and the environment in the development of human traits (see Nelkin and Lindee, 1995). In short, the distinction between genetic and non-genetic tests may not have justification.

Second, and picking up on a latter point, Juengst's taxonomy involves a hierarchy of kinds of genetic prediction that assumes a distinction between single-gene and complex disorders that may increasingly be unsustainable. In his taxonomy, single-gene disorders (e.g., fragile X syndrome, Huntington disease, PKU) provide the standard of investigation, testing, and treatment. Claims in favor of single-gene disorders include: (1) Because of the salience of one gene, single-gene disorders may be expressed regardless of the environment. (2) In single gene disorders, the products of one locus override the effects of products of the other loci. A consequence of the foregoing is that in single-gene disorders, (3) it is easier to learn the relationship between a gene and its phenotype. Since a gene product in a single-gene disorder is so deficient or dysfunctional as to disrupt homeostasis in significant ways, the single-gene disorder (4) is rare and (5) appears early in life. Finally, (6) most single-gene disorders resist treatment or management. (McInerney, 2002)

Certainly the distinction between single gene and complex diseases appears to be at the heart of genetics and our current understanding of genetic disease. Think for a moment about the typical high-school biology class, and the emphasis on single-gene disorders in the chapter on genetics. Now think that most individuals in the U.S. and Canada achieve at best this level of biology, absent the 861,000 or so who take the General Education Development (or high-school equivalency) Test (GED) (Kleiner, 2001). Yet, there is increasing evidence to show that most, if not all, genetic disorders are complex. Landers and Schork, for example, claim that "the category of complex traits is all-inclusive. Even the simplest genetic disease is complex, when looked at closely" (1994, p. 2037). As an illustration, sickle-cell anemia is typically given as a classic example of a simple Mendelian recessive trait. Yet, individuals carrying identical alleles at the β-globin locus can show markedly different clinical courses, ranging from early childhood mortality to a virtually unrecognized condition at age 50, thereby challenging point 5 above. The trait of severe sickle cell anemia is influenced by multiple genetic factors including a mapped X-linked locus and an inferred autosomal locus that can increase fetal hemoglobin amounts and thereby partially ameliorate the disease (Landers and Schork, 1994, p. 2037; also see Scriver and Waters, 1999), thereby challenging points 1,2, and 3 above. Beckwith and Alper echo this analysis: "there are no genes for a particular disease" (1998, p. 208). Even supposedly simple single-gene diseases, such as cystic fibrosis and Huntington disease, which were thought to be completely predictable on the basis or absence of a particular gene, display remarkable complexity. It is now known that the same altered CF gene can result in a wide variety of symptoms and that some people with the altered Huntington gene do not develop the disease at all (Beckwith and Alper, 1998, p. 208). Finally, that single-gene disorders are rare (4) and resistant to treatment (6) does not distinguish them in any essential way from complex genetic disorders. We can expect to find numerous complex genetic disorders that are rare and resistant to treatment.

    Third, the very concept of the "gene" is ambiguous. The complexity of the concept of the gene, a term coined by the Danish botanist Wilhelm Johannsen (1909)[9], is in part due to its evolving character (Portin, 1993, 2002; Moss, 2003; Biological Sciences Curriculum Study, 1997). Our modern understanding of the gene is constructed around an ever-growing collection of fundamental commitments. Consider, for example, traditional or Mendelian concepts of inheritance: (1) alleles segregate, (2) non-alleles assort independently, (3) traits can show a dominant or recessive pattern of inheritance, (4) genotypes give rise to phenotype, (5) chromosomes carry hereditary information, (6) DNA is the informational component of chromosomes, (7) inheritance is nuclear and vertical, (8) the genetic contribution from each parent is equal, (9) genes are units of inheritance, (10) genes occur at fixed locations, (11) genes on the same chromosome are linked, and (12) the laws of probability help to explain patterns of inheritance (Biological Sciences Curriculum Study, 1997, p. 36). Nevertheless, science has uncovered during the last few decades numerous exceptions reflecting the diversity of living systems.

    Consider, in contrast, a non-traditional or Post-Mendelian understanding of the gene: (1) some heritable traits are extranuclear, (2) genetic anticipation correlates with expansion of trinucleotide repeats, (3) genomic imprinting can alter the expression of genetic information and distinguish its parental origin, (4) both chromosomes of a pair may, in rare cases, come from one parent, (5) some genes are mobile and insert themselves in new chromosomal locations, (6) genes may, in rare cases, undergo horizontal transfer between individuals or species, (7) many traits result from expression of more than one gene combined with environmental factors (multifactorial inheritance), and (8) genetic information specified by genomic sequence may be altered during RNA editing (Biological Sciences Curriculum Study, 1997, p. 43).

    The message here is that the concept of the gene changes. Portin (1993) suggests that we think of the change in the concept of the gene in terms of three distinct periods of thought. First, there is the *classical view* (early 1910 to early 1940) in which the gene "was widely regarded as an indivisible unit of genetic transmission, recombination, mutation, and function; all of these criteria of the gene led to one and the same unit of genetic material" (Portin, 1993, p. 175). In other words, the action of one gene is seen to lead to one enzyme, and so on. Second, there is the *neoclassical view* (late 1940s to early 1970s) in which the gene is a cistron, or stretch of DNA made up of contiguous nucleotides, which code for a single-messenger RNA and thus indirectly for a single polypeptide (or building block of protein) (Watson and Crick, 1953; Watson, 1968). In other words, the action of one cistron is seen to lead to one polypeptide. Third, there is the *modern period* (late 1970s to present) in which differing views emerge. On the one hand, the gene is a physical entity or "element" (Hartl et al., 2002) with particular functions and may be investigated on the molecular level (Watson and Crick, 1953; Watson, 1968). On the other hand, the gene is a unit of transmission that may be investigated using the tools of evolutionary biology (Mayr, 1982, 1988; Cavelli-Sforza and Feldman, 1981). Between these two accounts, so to speak, there are numerous complications reflecting the diversity of living systems (Carlson, 1991, p. 477).

The lessons of the foregoing are that the distinctions between genetic and non-genetic tests, and single and complex genetic disorders, are not that clear. Further, the concept of the gene is evolving.

2.2. Implications

How, then, do we understand genetic disease?   We could begin by recognizing that knowledge of the structure and function of the genetic material has outgrown the terminology traditionally used to describe it. It is arguable that the old term "gene," essential at an earlier stage of the analysis, is no longer useful, except as a handy and versatile expression, the meaning of which is determined by the context. In that case, the challenge is to devise a new terminology for use when precision is needed (Portin, 1993, p. 208).

As a start, one might take a closer look at concepts such as "inheritance" (which emphasizes process as opposed to thing), "variation" (which emphasizes the dynamic character of inheritance as opposed to a static state of causal elements), and "individuality" (which emphasizes locality or local expression as opposed to typology or generic classification). To begin with, inheritance is a complex idea. Inheritance concerns the biological roots and processes of that which is transferable or descendible from an ancestor. In that it concerns the biological components of the continuation of a species, it involves a study of basic components of such continuations, such as DNA, RNA, mitochondria, and proteins. In that it studies that which is being transferred and how within a species over time, it concerns the environments of the biological components. In the end, it concerns both biology and environment. One of the advantages of using the term inheritance is that the necessary relation between gene and environment, or nature and nurture, is emphasized.

Further, inheritance is the study of biological variation within environments. Now that the HGP has spelled out the 3 or so billion chemical letters of the average human's DNA code, the hunt is on for what researchers call single nucleotide polymorphisms (SNPs), the locations along the DNA chain where spelling differs from one person to the next. Humans are remarkably similar: 999 out of every 1,000 letters are identical among all people. Yet the genetic code is so long that those one-in-a-thousand variations add up to make some people tall, some curly haired, others prone to heart attack, and others resistant to HIV. Even among individuals within the same family, SNPs can vary and lead to differences in phenotypic expression. One of the advantages of using the term variation is that biological complexity is emphasized.

Put another way, inheritance is the study of biological individuality. In all, there are probably ten million SNPs, and decoding how they affect individuals could revolutionize both the diagnosis and treatment of disease. Consider, for example, that studies on asthma patients show variations in how patients respond to the drug Albuterol. The drug Albuterol affects the so-called (beta) 2 adrenergic receptor, which in turn affects lung function. There are thirteen different known SNP areas on this gene, which theoretically could be combined to form 8,192 different SNP patterns (called "halotypes") (Fischer, 2000). One of the advantages of the term individuality is that the potential for variation in any part of the human genome and individuality in expression of human health and disease is emphasized.

Given a change in terminology, we may begin to address the concerns raised by many critics of the genetic revolution (e.g., Lewontin, 1991, p. 17ff; Childs and Scriver, 1986; Childs, 1994) regarding the misuse of the term "gene" and its attendant metaphors such as "the gene for" and "it's in the genes" (which is deterministic)[10] and "it's all in the genes" (which is fatalistic). Such representations of the gene would be less available in the literature because the speaker would have to make precise statements regarding the relations between and among that which is bringing about what. Yet, we would also support the evidence (e.g., Landers and Schork, 1994) that there *are* causal relations between biological preconditions and outcomes that can be substantiated. These events simply have to be substantiated carefully.

3. BIOETHICAL CHALLENGES

Improved understanding of genetic and environmental contributions to complex disease should shift the focus of medicine from the name of the disease itself to genetic individuality and to the individuality of the experiences, habits, and conditions of the environment of the particular patient, for purposes of survival. With this shift comes a host of bioethical issues.

To begin with, an individualized or localized concept of genetic disease raises questions concerning the legitimacy of any single set of standards or classificatory scheme of disease in medicine. As this chapter illustrates, differences in how genes are expressed, in how genes are influenced by their environment, and in how cultures interpret such expressions, must be taken into consideration in framing disease categories (Spector, 1966; Osborne, 2001). Differences turn on how normality and abnormality are understood. Here normal can mean (Murphy, 1976, pp. 117-133): (1) statistical, (2) average or mean, (3) typical or expectable, (4) conducive to the survival of the species, (5) innocuous or harmless, (6) commonly aspired to, and (7) most perfect or excellent of its class. Genetics shows us that pathological processes are normal (in terms of 1,2, 3, and especially 4). In this sense, it is normal to be diseased. Alternatively, genetics cannot convince many of us that it is good to be diseased. In this way, disease stands in contrast to that which is normal (in terms of 5, 6, and 7) and thus designated abnormal. The point here is that the relation between normal and abnormal, or health and disease, is complex, an insight already established by Lewontin (1991).

This complexity has implications for treatment, leading us to question the legitimacy of any single type of standard of treatment for patients with a particular genetic disorder. At one end of the spectrum are genetic tests intended to identify people at increased risk for the disease and recognize genotype differences that have implications for effective treatment. Pharmacogenomics focuses on crucial differences that cause drugs to work well in some people and not at all, or with dangerous adverse reactions in others. For example, researchers investigating Alzheimer's disease have found that the way patients respond to drug treatment can depend on which of three genetic variants of the *ApoE* (Apolipoprotein E) gene a person carries. Increasingly, we can expect health care professionals to use genetic tests to match drugs to an individual patient's body chemistry. Correspondingly, we can expect pharmaceutical companies to develop highly specific drugs for individuals with particular conditions (Fischer, 2000).

At the other end of the spectrum are new drugs and therapies that specifically target the biochemical mechanisms that underlie the disease symptoms or even replace, manipulate, or supplement nonfunctional genes with functional ones. Collectively referred to as gene therapy, these strategies typically involve adding a copy of the normal variant of a disease-related gene to a patient's cells. The most familiar examples of this type of gene therapy are cases in which researchers use a vector to introduce the normal variant of a disease-related gene into a patient's cells and then return those cells to the patient's body to provide the function that was missing. This strategy was used in the early 1990s to introduce the normal allele of the adenosine deanimase (ADA) gene into the body of a little girl who has been born with ADA deficiency (National Coalition, 2002). In this disease, an abnormal variant of the ADA gene fails to make adenosine deanimase, a protein that is required for the correct functioning of T-lymphocytes. Other approaches to gene therapy include correcting genetic defects that involve only a single base change (chimeraplasty) and introducing new genetic information into cells leading to genetic defects.

The bioethical challenges raised by a localized account of genetic disease are numerous. Depending on how distinctions between normality and abnormality are drawn, individuals will be regarded as diseased or healthy and the investment of health care resources will be seen to be necessary or unnecessary. As Chapter 4 establishes, disease nosologies function as treatment warrants, in this case, how we understand genetic disease sets up the justifications for treatment. In the case of genetics, treatment options are complex. Treating everyone, in a world of limited resources and genetic knowledge, is impossible. Treating everyone is, according to population geneticists, contraindicated given the role mutations play in the adaptation of a species to environment. Treating anyone with germ-line therapy is considered by some a violation of natural law (Coors, 2002) or sanctity of life. Treating a select group of individuals raises issues concerned with just access to health care. Then there are those who will choose not to be treated even in the face of recommendations to be treated. Limits on treatment will turn, then, on the availability of resources, on what constitutes known and appropriate foci of intervention, on the criteria employed, and on the wishes of patients. The bioethical issues of utility (i.e., benefit/burden calculations), informed consent, patient autonomy, and justice become central in discussions of how we understand and treat disease in an era of genetic medicine.

4. NEGOTIATING DIVERSE VALUES IN A PLURALIST SOCIETY

Essentially advances in medical sciences are causing information from genetic tests to be more useful in the diagnosis and treatment of human disease and illness. An implication of this is that some people are discouraged from participating in genetic testing and other means of genetic diagnosis because of the potential that insurers would use that information to deny or reduce health insurance coverage. In response to this concern, numerous states (Cutter, 1998; Hudson et al., 1995) have engaged in debates regarding the extent to which the insurance industry ought to have access to genetic information. The Colorado Statute[11], for example, defines genetic testing as a direct laboratory test of human DNA, RNA, or chromosomes used to identity the presence or absence of alterations in genetic material associated with illness or disease. The statute applies to entities that provide health, group, disability, and long-term insurance and are within the Colorado Insurance Commission's jurisdiction. The covered entities are prohibited from seeking, using, or keeping genetic information for underwriting or nontherapeutic purposes. Violation of the act is an unfair insurance practice subject to Insurance Commission sanctions. The statute provides a private right of action for individual injured by wrongful use of genetic information, with both legal and equitable remedies available. Additionally, the prevailing party may recover attorney fees.

The development of the Colorado Statute limiting access to genetic information by insurers illustrates an attempt to negotiate differing values regarding a medical event in a pluralist society. For instance, encouraging patients to take genetic tests is a good because it promotes the interests of the patient and related parties, particularly in preventive, palliative, or curative treatment can be offered. Preventing insurer's access to patient genetic testing information is good insofar as insurers would be unable to cherry-pick their client pool. For some, there is a right to personal genetic information. Yet, allowing insurers access to genetic testing information can also be considered a good insofar as the information is used to provide actuarial data on emerging clinical problems that need attention and therefore coverage by current and potential clients. Access also allows insurers to develop risk classifications, which promotes the financial soundness of a voluntary insurance market. For some, businesses have a right to practice responsibly. One notes here a tension between differing senses of the good and right and their goals.

How one begins to resolve these differences is anything but simple. In the development of public policy, which medicine increasingly find itself, there appears to be four components, together which provide guidance in resolving value disputes. First is access. Interest parties must have the opportunity to enter into debates. In the case of developing legislation in Colorado, numerous groups participated in discussions about regulating access and use of genetic information by the insurance industry. These groups included, among others, patients, insurers, lawyers, governmental agencies, geneticists, physicians, philosophers, theologians, members of interests groups, and the media. Numerous and well publicized meetings make access possible.

Second is information. The goal here is to increase the understanding of participants so that discussions may occur. In the case of developing legislation in Colorado, meetings entailed numerous educational presentations to ensure that participants understood the science, technology, ethics, business, and law of storing genetic information. Different sides had ample opportunity to present their information and case.

Third is noncoercion. Failure to provide relevant information, deception threats, and outright threats of violence count as coercive tactics and ought to be avoided, as was the case in developing legislation in Colorado. However, one will need to distinguish between acts of coercion and those of peaceful manipulation. If one understands coercive actions as those that place or threaten to place a person in a disadvantaged state without justification, and if one defines peaceful manipulation as those actions that place or offer to place a person in an advantaged state to which the person is entitled, coercions will be forbidden and peaceful manipulations will be allowed. One can say that the development of any legislation will involve these kinds of forces. One would hope that the latter is used as opposed to the former.

Fourth is compromise. In the case of Colorado legislation, supporters of the bill met privately on numerous occasions with members of the insurance industry and other agencies (e.g., enforcement) to clarify and settle disagreements prior to the introduction of the bill to the Colorado legislature. Disagreement centered on (1) how to define genetic testing (and the related issue, genetic disease or illness), (2) what entities to regulate, (3) what entities to exempt, and (4) what sanctions to mandate. What became clear during discussions is that, despite their significant differences, parties were able to establish a shared goal, namely, to reform health care insurance. While members of the insurance industry held that reform was needed in order to develop a level playing field, others wished to provide access to genetic testing without fear of losing health insurance coverage. This shared goal of reform allowed the possibility of continued discussions regarding not *whether* the state should regulate but rather *how* the state should regulate access to genetic information by the insurance industry.

In short, negotiation involving matters of public policy, in this case health policy, entails providing interested parties an opportunity to access discussions, to receive sufficient information, to be protected against coercive forces, and to compromise. The parallels to the foregoing discussion about the development of our understanding of disease are striking. To begin with, the naming and definition of disease (in this case genetic) and its tests are not trivial and have important consequences. Differences in interpretation turn on the individual interests and the goals that are sought. In this case, most patients see disease labels as personal information while insurers rely heavily on such labels to run a business. Tensions between private and public goals arise. In addition, negotiation is seen to play a key role in developing legislation regarding access to genetic information. This is not unlike the negotiation that takes place when clinical nosographies and nosologies are developed (e.g., in the case of AIDS, Chapter 2).

5. CLOSING

Given current advances in biology, it is surely the case that we will need to reflect on our moves in genetic science and medicine (Cutter, 2002b). And if the popular literature is any indication, it is not too early to worry. When it comes to talking about disease in an era of genetic medicine, much depends on how we understand the gene. When it comes to talking about the gene, we are advised to "proceed with caution" (1989). The foregoing is offered as a cautionary measure in our discussions and practice of genetic medicine. It is offered as well as an illustration of the importance of thinking through how disease, in this case genetic disease, is understood.

CHAPTER ENDNOTES

1. This essay was originally presented to students in PHIL 320, "Philosophy of Science" in Spring 1997. I am indebted to my colleague, Rex Welshon, Department of Philosophy, for encouraging me to put these ideas into words. I am grateful as well to Mike Dougherty, Department of Biology, Hampden-Sydney College, Virginia, for his comments on the chapter.

2. The number has been a source of debate. In February 2001, scientists announced that humans probably have fewer than 35,000 genes (as opposed to the earlier estimation of 100,000 genes). This compares to the 19,000 that a C. elegans worm has and 50,000 that rice has (Weiss, p. A1).

3. The language of clinical problems in genetics is ambiguous and will not be pursued here. Distinctions among disease, illness, disorder, dysfunction, and defect are important and worth of further consideration (Cutter, 1993).

4. Determinism is the doctrine that every event has a cause. The main interest in determinism has been in assessing its implications for free will. Here human action can be effectual, albeit in small amounts.

5. Fatalism is the doctrine that human action has no influence on events.

6. Juengst recognizes that the categories may overlap and tests could fall into several categories. For example, a Huntington disease mutation test could be diagnostic (category 1) if it were used to rule out a diagnosis of Huntington disease in a neurologically impaired patient. Or, it could be prognostic (category 2) based on genetic anticipation, i.e., earlier onset or increased severity of an inherited disorder in subsequent generations.

7. The term "prophylactic" emphasizes "that there are interventions to be made, that the problem is not internal or inevitable" (Juengst, 1995a, p. 33).

8. Penetrance refers to the proportion of organisms whose phenotype matches their genotype for a given trait. A genotype that is always expressed has a penetrance of 100%. A penetrance of less than 100% (incomplete penetrance) is the extreme of variable expressivity in which the genotype is not expressed to any detectable degree in some individuals. For example, a person with a genetic predisposition to lung cancer may not get the disease if he or she does not smoke tobacco. A lack of gene expression may result from environmental conditions, such as in the example of not smoking, or from the effects of other genes (Hartl, 2002).

9. Johannsen also coined the terms "genotype" and "phenotype." He did so precisely as a critique of preformationist fallacies and on behalf of a return to holism defined in terms of the full range of developmental phenotypic potentials associated with any genotype (Moss, 2003, pp. 28ff).

10. Consider the August 1994 cover of *U.S. News and World Report*, which declares: "Infidelity: It May Be In Your Genes".

11. The Colorado Statute Limiting Access to Genetic Information is as follows:
10-3-1104-7. GENETIC testing–declaration–definitions–limitations on disclosure of information–liability–legislative declaration.
(1) The general assembly hereby finds and determines that recent advances in genetic science have led to improvements in the diagnosis, treatment, and understanding of a significant number of human diseases. The general assembly further declares that:
    (a) Genetic information is the unique property of the individual to whom the information pertains;
    (b) Any information concerning an individual obtained through the use of genetic techniques may be subject to abuses if disclosed to unauthorized third parties without the willing consent of the individual to whom the information pertains;
    (c) To protect individual privacy and to preserve individual autonomy with regard to the individual's genetic information, it is appropriate to limit the use and availability of information;
    (d) The intent of this statute is to prevent information derived from genetic testing from being used to deny access to health insurance, group disability insurance, or long-term care insurance coverage.
(2) For purposes of this section:
    (a) "Entity" means any sickness and accident insurance company, health maintenance company , nonprofit hospital, medical-surgical and health service corporation, or other entity that provides health care insurance, group disability insurance, or long-term insurance coverage, and is subject to the jurisdiction of the commissioner of insurance.
    (b) "Genetic testing" means any laboratory test of human DNA, RNA, or chromosomes that is used to identify the presence or absence of alterations in genetic material which are associated with disease or illness. "Genetic testing" includes only such tests as are direct measures of such alterations rather than indirect manifestations thereof.
(3)   (a) Information derived from genetic testing shall be confidential and privileged. Any release, for purposes other than diagnosis, treatment, or therapy, of genetic testing information that identifies the person tested with the test results released requires specific written consent by the person tested.
    (b) Any entity that receives information derived from genetic testing may not seek, use, or keep the information for any nontherapeutic purpose or for any underwriting purpose connected with the provision of health care insurance, group disability insurance, or long-term care insurance coverage.
(4) Notwithstanding the provisions of subsection (3) of this section, in the course of a criminal investigation or a criminal prosecution, and to the extent allowed under the federal or state constitution, any peace officer, district attorney, or assistant attorney general, or a designee thereof, may obtain information derived from genetic testing regarding the identity of any individual who is the subject of the criminal investigation or prosecution for use exclusively in the criminal investigation or prosecution without the consent of the individual being tested.
(5) Notwithstanding the provisions of subsection (3) of this section, any research facility may use the information derived from genetic testing for scientific research purposes so long as the identity of any individual to whom the party pertains is not disclosed to any third party; except that the individual's identity may be disclosed to the individual's physician if the individual consents to such disclosure in writing.
(6) This section does not limit the authority of a court or any party to a parentage proceeding to use information obtained from genetic testing for purposes of determining parentage pursuant to section 13-25-126, C.R.S.
(7) This section does not limit the authority of a court or any party to a proceeding that is subject to the limitations of part 5 of article 64 of title 13, C.R.S., to use information obtained from genetic testing for purposes of determining the cause of damage or injury.
(8) This section does not limit the authority of the state board of parole to require any offender who is involved in a sexual assault to submit to blood tests and to retain the results of such tests on file as authorized under section 17-2-201 (5) (g), C.R.S.
(9) This section does not limit the authority granted the state department of health and environment, the state board of health, or local departments of health pursuant to section 25-1-122, C.R.S.
(10) This section does not apply to the provision of life insurance or individual disability insurance.

(11) Any violation of this section is an "unfair practice," as defined in section 10-3-1104 (l), and is subject to the provisions of sections 10-3-1106 to 10-3-1113.

(12) Any individual who is injured by an entity's violation of this section may recover in a court of competent jurisdiction the following remedies:

(a) Equitable relief, which may include a retroactive order, directing the entity to provide health insurance, group disability insurance, or long-term care insurance coverage, whichever is appropriate, to the injured individual under the same terms and conditions as would have applied had the violation not occurred; and

(b) An amount equal to any actual damages suffered by the individual as a result of the violation.

(13) The prevailing party in an action under this section may recover costs and reasonable attorney fees (Colorado Revised Statute, 10-3-1104.7).

CHAPTER 10

CONCEPTS OF GENDERED DISEASE[1]

Much has been said in the latter part of the twentieth century about the importance of addressing more closely women's disease and health.   Discussions have led to an increase of women physicians, women's health care centers, and the involvement of women in clinical research projects. The call is for more attention on women's disease and health in order to better serve women. In other words, we are called to pay closer attention to how gender frames disease. This chapter takes a closer look at the role gender plays in the construction of disease, thereby drawing upon the analysis provided earlier on the social construction of disease, particularly Chapters 6, 7, and 8. It argues that a non-neutral, or rather a non-gender-neutral, account of disease is non-defensible. The task, then, is to determine the kind, level, or degree of gender bias in the framing of women's disease that is appropriate for judgment and action.

1. THE MEDICALIZATION OF WOMEN'S PLACE:
SOME NOTES FROM HISTORY

Since at least the time of Hippocrates (approx. 460-377 B.C.) (*Regimen,* 1943) and Aristotle (384-322 B.C.) (*The Generation of Animals,* 1984, 726ff), medicine has justified the boundaries, nature, and treatment of women's clinical problems in remarkably consistent ways (Tuana 1988). In earlier chapters, we learned that in framing disease, the descriptive level interplays with the prescriptive level in particular contexts. Put another way, "is" and "ought" interplay. We see this played out in medicine particularly with regard to how women and their conditions are medicalized. Consider briefly these few notes from the history of Western medicine.

First, woman is *not* like man and is in fact "a sort of natural deficiency" (Aristotle, quoted in Tuana 1988, p. 41) or "less perfect" (Galen, quoted in Tuana, 1988, p. 42) compared to the male. Physically, she is frailer, her skull is smaller, and her muscles are more delicate. Even more striking is the difference between the nervous system of the two sexes.   The female nervous system is prone to overstimulation, resulting in irritability, hysteria, and ultimately exhaustion. "The female sex," as Dr. Hall explained in 1827, "is far more sensitive and susceptible than the male, and extremely liable to those distressing affections which for want of some better term, have been denominated nervous..." (quoted in Smith-Rosenberg and Rosenberg, 1981, p. 283).

Second, woman *is* her reproductive organs. As Dr. Holbrook put it in 1882, it was "as if the Almighty, in creating the female sex, had taken the uterus and built up a women around it" (quoted in Smith-Rosenberg and Rosenberg 1981, p. 284). She is her uterus; she is a body that functions in a particular way for purposes of reproduction. She is a reproducing machine. Here a mechanistic model of body supported by René Descartes (1596-1650) and Julien Offray de la Metrie (1709-1751) (Robinson, 1973, Ch. 1) finds expression and is applied to women and their roles as producer and mother.

Third, woman *should* reproduce. Because woman is by nature a reproductive organism, she should live in accordance with her nature to reproduce. It is "natural" that woman display more affect, emotions, sympathies, and passions than man (see, e.g., Freud 1962 [1905], III). It is "natural" that woman exhibits less desire for the spiritual life (Butler, quoted in Curren, 1991). And it is "natural" that woman express little interest in education (Potter, 1891, quoted in Smith-Rosenberg and Rosenberg, 1981, p. 285; Clouston, 1882, quoted in Smith-Rosenberg and Rosenberg, 1981, p. 288). The view is that woman's natural constitution causes her to behave in certain compromised or faulty ways.

Fourth, woman *should* not deviate from their nature. Regarding her diseased status, the intimate link between the reproductive system and the nervous system is the basis for the "reflex irritation" model of disease causation so popular in middle- and late-nineteenth-century texts and monographs in psychiatry and gynecology. Any imbalance, exhaustion, infection, or other disorder of the reproductive organs leads to pathological reactions in other parts of the body, with the result that changes in the reproductive cycle shape emotional states. As Dr. Dirix put it in 1856, "These diseases will be found, on due investigation, to be in reality, no disease at all, but merely the sympathetic reaction or the symptoms of one disease, namely, a disease of the womb" (quoted in Smith-Rosenberg and Rosenberg, 1981, p. 284). Note Dr. Rohe's analysis of "lactational insanity" (i.e., depression) at the 44th Annual Meeting of the American Medical Society (1893):

> Prolonged or excessive lactation is given as the chief cause [of insanity (i.e., depression)] ....
> In most cases, this is probably true, yet there are some cases in which the disease must be attributed to other etiological factors.... It appears...that the local [i.e., pelvic] irritations acting upon the central organ [i.e., brain] are active, both as determining the duration as well as the course of the mental disorder (quoted in Weissman and Olfson, 1995, p. 799).

The point is that woman's deficient constitution causes her diseased state. She should not deviate from her nature lest she is prepared to be diseased.

Fifth, socially speaking, woman is *destined* to be patient. Physicians see woman as the product and prisoner of her reproductive organs. Woman's uterus and ovaries control her body and behavior from puberty through menopause. As Dr. Engelmann, President of the American Gynecological Society, observed in 1900:

> Many a young life is battered and forever crippled in the breakers of puberty; if it crosses these unharmed and is not dashed to pieces on the rock of childbirth; it may still ground on the ever-recurring shallows of menstruation, and easily, upon the final bar of the menopause ere protection is found in the unruffled waters of the harbor beyond the reach of sexual storms (quoted in Smith-Rosenberg and Rosenberg 1981, p. 285).

Woman is indeed a pitiful thing.

On this account, then, woman's problem is that she is not like him, she is different, and difference is deficit (Tavris 1992). Her problem is that she is a slave to her reproductive organs. Her problem is that she is destined to be patient. Her problem becomes his problem as he the doctor interprets it (Achterberg 1991). And her problems cannot be treated because she is destined to be patient.

2. RISE OF THE U.S. WOMEN'S HEALTH MOVEMENT

The rise of the women's health movement in the 1970s is a result of a complex set of influences. Issues raised by advocates of the rights of minorities, consumers, the mentally ill, and prisoners often include health care components and help reinforce public acceptance of acknowledging women's concerns regarding their health care. The women's health movement in particular encourages women to question established medical authority, to take responsibility for their own bodies (Boston Women's Health Book Collective, 1973), and to express new demands for clinical research and access to appropriate health care.

Between 1974 and 1983, the National Commission for the Protection of Human Subjects of Biomedical and Behavioral Research (1978) and the President's Commission for the Study of Ethical Problems in Medicine and Biomedical and Behavioral Research (1983) develop guidelines that require any research project that is federally funded to ensure humane treatment for human subjects (both females and males), including the acquisition of informed consent. In 1985, a Task Force on Women's Health Issues begins work to aid the Public Health Service (PHS) "to improve the health and well-being of women in the United States" (Department of Health and Human Services, 1985). In 1993, the National Institutes of Health (NIH) (1993) issue guidelines to ensure that federally funded investigations include an analysis to determine whether the interventions being studied affect women and members of minority groups differently from other groups. Further, section 429B of the NIH Revitalization Act enjoins the NIH Director to ensure that women and members of minority groups are included in all research projects, unless exclusion is appropriate because of health, the specific focus of the research, or other circumstances that the NIH Director approves. In 1993, the U.S. Food and Drug Administration (FDA) (1993) issues guidelines concerning the participation of women in studies of medical products. Guidelines state that scientists must formulate research hypotheses so as not to exclude sex as a crucial part of the research question being asked. For example, when exploring the metabolism of a particular drug, one must routinely run tests on both males and females. The goal is to analyze potential differences in drug reaction and efficacy between the sexes. In 1993 as well, the FDA alters a 16-year-old policy that had excluded most women of child-bearing potential from the early phases of clinical trials. And, in 1995, as reports from the Fourth World Conference on Women in Beijing and an issue of *Science* (Vol. 269, August 11, 1995) illustrate, women's health and disease emerge as foci of concern for researchers, health care practitioners, organizations, institutions, and governments in the global order. Such attention continues into the Third Millennium with added emphasis on woman and HIV/AIDS and other immune deficiency conditions, the efficacy and safety of hormone replacement therapy, safe motherhood practices, and heart disease (World Health Association, 2003).

Crucial to developments involving interest in women's health is the recognition that, among females and males, many diseases have different frequencies (e.g., depression, lupus, anorexia nervosa), different symptoms (e.g., AIDS, heart disease, alcoholism, gonorrhea), and different complications (e.g., heart disease, connective tissue disease, sexually transmitted diseases). These differences can no longer be seen as inherent deficits or deficiencies (Tavris, 1992). Rather, they point to significant variations in biological and psychosocial expressions, a theme taken up at great length in previous chapters, and especially Chapter 9. Such differences call for varying understandings of particular diseases and modes of intervention, including preventive, palliative, curative, and long-term care.

In addition, previously-accepted assumptions and practices are challenged. The so-called moral responsibility to protect the reproductive capacity of women and fetuses is seen to be paternalistic representing unwarranted institutional intervention (e.g., *Automobile Workers v. Johnson Controls*, 111 S Ct. 1196, 1991; Boston Women's Health Collective, 1998, p. 148). Previously held justifications for excluding women from research (e.g., women's hormonal complications, women's special duties to the unborn and future generations, and women's failure to conform to an identifiable convenient cohort [e.g., male veterans]) are seen to be indefensible.

Then there is increased public awareness and public outcry, particularly by women. The public is outraged that results from large scientific studies involving men are applied to women. For instance, the largest study investigating the effects of aspirin on heart disease rates involves exclusively men; yet the results continue to be used to advise women (Dresser, 1992, p. 24). The presumptions are that heart disease is expressed identically in men and women, that the appropriate mode of treatment will be same, and that all men and all women are the same, all of which is not the case.

As a consequence, there emerges a growing consensus that women (1) need to be listened to more closely, (2) need to be included in research projects, and (3) require their own special studies (Dresser, 1996; Rosser, 1992; Merton, 1996) and clinical settings (Lurie, 1993; Purdy, 1996)[2]. Further reflections conclude that any consideration of the role gender plays in the construction of disease must be accompanied by ones recognizing the interplay among gender, ethnicity, and class.[3] Such would bring about a broader view of women's health and disease, one that goes beyond the confines of the reproductive system (Achterberg, 1991; Borysenko, 1996; Wolf, 1996, Mahowald, 1999) and leads to more appropriate definitions of women's disease and consequently better health care.

3. GENDER MATTERS

One of the messages of late twentieth century medicine is that gender matters.[4] On first response, the recognition that gender matters appears adventitious to the promotion of women's health. Singling out women engaging in careful studies appears to lead only to greater knowledge about women's health and disease and thus better clinical care. We now know that the X chromosome has 164 million bases whereas the Y chromosome has 59 million bases, leading to work on important differences between males and females in how they express disease, especially X-linked disorders (e.g., fragile-X, hemophilia). Singling out women is an appropriate response to past discriminatory practices in medicine. Prior to the rise of the women's health movement, clinical problems are seen to be a consequence of women's difference with men, and difference is deficit (Tavris, 1992). Women's problems (e.g., depression, insanity) result from their reproductive organs (Weissman and Olfson, 1995; Smith-Rosenberg and Rosenberg, 1981), which destine women to be patients. It's time to address these false assumptions.

Yet, on second thought, singling out women may lead to a segregation that might be harmful. Women share many biological features with men, and are, genetically speaking, 99.9% the same. A separation between men and women's health research may put women's research behind and delay the development of possible treatments. Alternatively, forced inclusion of women in medical research may lead to unfunded projects, given that study sizes might have to be larger in order to accommodate women as subjects (Meinert, 1995). Studies of women from particular racial, ethnic, and class groups could indicate that these populations are more susceptible to conditions that reinforce group stereotypes. The question is, then, how can we appropriately accommodate gender into clinical nosology and nosography?

To begin with, in medicine, *sex*[5] (L. *sexus*, sex) matters. The distinction between female and male, XX and XY, is one that dates back to the beginning of medicine[6] and provides a basis for concerns about diseases particular to females and males. There are diseases particular to organs that are found exclusively in one sex (e.g., ovarian or uterine cancer, and penal or prostrate cancer). There are diseases particular to females (e.g., fragile X) and males (e.g., hemophilia) that have to do with the sex chromosomes, something that we are beginning to understand even more today given the rise of genetics.

Yet, we need to recognize that the distinction between and among sex differences is neither exhaustive nor exclusive. Human beings display a range of sexual diversity. There are those who fall in so-called exclusive genetic classes (e.g., a "normal" homologous XX [female] or nonhomologous XY [male]). There are those who do not fall within either class (e.g., those with Turner's syndrome or Klinefelter's syndrome) and those who fall in both (e.g., hermaphrodite[7], male pseudohermaphrodite[8], and female pseudohermaphrodite[9]). Studies show at least 9% of the human population express sexual diversity (Fausto-Sterling, 1987). In short, sex is a malleable continuum that challenges the constraints of binomial categories (Hubbard, 1995; Fausto-Sterling, 1992, 1996, 2000; Intersex Society of North America website, www.isna.org).

Nevertheless, there is something that we can say about sex differences or distinct expressions of sexuality. Anything does not go. There are limited expressions of genetic sexuality (e.g., XX, XY, XO, XXX, XXY, XYY, XXXY), gonadal sexuality (e.g., ovaries, testes), hormonal sexuality (e.g., presence of estrogens or androgens), genital sexuality (e.g., clitoris, penis), sex assignment (e.g., "The baby is female!" "The baby is male!" "The baby is trans-sexual."), and sex identity (e.g., "I am female." "I am male." "I am gay." "I am am lesbian." "I am trans-sexual."). Limited expressions of genetic sexuality are in part a function of select ways to propagate the species[10]. These limitations and expressions of sexuality come about through various methodological assumptions including judgments concerning simplicity, orderliness, usefulness, and predictability. They support accepted divisions in medicine which are in part a result of work in medical specialties such as obstetrics, gynecology, reproductive biology, urology, and proctology. They give rise to gender classifications such as feminine and masculine, girl and boy, and woman and man, trans-sexual and inter-sexual, not to mention the pejorative ones. Clinical medicine and the classifications it designs presuppose and make empirically explicit sexuality and gender in various and complex ways.

Classifications of sexuality give rise to the development of norms, and particularly to *gender*-norms. Here gender (L. *genus*, race) refers to a social category of sexuality. In contexts where sexual or gender roles are well-entrenched, the corresponding norms function prescriptively: they serve as the basis for judgments about how individuals ought to be, act, and so on. Furthermore, we decide how to act, what to strive for, and what to resist in light of such norms. Gender-norms of femininity and masculinity are clusters of characteristics and abilities that function as standards by which individuals are judged to be "good" instances of their gender. On one prominent model, for example, to be "good" at being female or feminine, one should be nurturing, emotional, cooperative, sexually restrained, pretty, etc; to be good at being a male or masculine, one should be strong, active, independent, rational, sexually aggressive, handsome, etc. (Achterberg, 1991, Part IV).[11]

As framed here, the descriptive level of analysis in discussions of sexuality and gender is inextricably tied to the prescriptive level. Sex is tied to gender, female to femininity, male to masculinity, trans-sexual to bisexual, etc. Facts and values, as Engelhardt (1996) and others (e.g., Fleck, 1979 [1935]; Wartofsky, 1976) teach us, interweave in complex ways. Observations in medicine are always ordered around theoretical assumptions, including judgments concerning how to select and organize evidence into explanations. Further, observations are always ordered around evaluative assumptions, including those concerning what objects are assigned significance and what actions are appropriate in order to achieve certain goals.

Moreover, the prescriptive force is backed by *social* sanctions fashioned in light of what goals are seen as worthy of achievement. If one aspires to conform to such norms, one is rewarded. This is the message forwarded, for example, by cosmetic surgeons who prescribe facial lifts and health care professionals who prescribe hormone therapies (Borysenko, 1996, Chs. 11-13; Northrup, 2001). If one does not conform, one is censured, sometime weakly and sometimes severely. For example, if a woman's behavior violates expected gender-role norms, her behavior is frequently attributed to various physical or mental illnesses and in turn treated in a variety of ways, including name calling (e.g., "noncompliant," "b..."), pharmaceutical agents (e.g., anti-depressants [Greenspan, 1993]) and gynecological surgeries (e.g., hysterectomies and clitorectomies [Waisberg and Paige, 1988; Broverman et al., 1970]).

In short, facts, values, and culture interplay in fashioning disease. Put another way, given that sex matters in the construction of disease, and gender is inextricably tied to sex, gender matters in the construction of disease.

4. GENDERING DISEASE

Given that gender matters in the construction of disease, the task is to determine the kind, level, or degree of gender bias that is justifiable in the construction of disease and create the conditions appropriate for judgment and action.

Feminists such as Susan Sherwin (1992) and Sue Rosser (1992) assert that the exclusion of women in research is a form of continued oppression and that such oppression must be voiced, criticized, resisted, and responded to with alternatives that promote women's emancipation. This direction leads us to seek to change the institution of medicine, to encourage women to enter into high-level professional roles, and to design policies that accommodate women in clinical and scientific medicine. Good strides have taken place on this level. Medicine already has begun to incorporate women into clinical research projects in order to study gender differences that might influence diagnostic and treatment procedures (National Institutes of Health, 1993). Women are clearly recognized as a significant consumer group in medicine (Dresser, 1996, p. 149). Women are increasingly being admitted to medical schools and promoted to higher ranks in medical administration (Achterberg, 1991, Part IV).

Such socio-political changes have been requisite for the reconstruction of medical institutions and the status of women as patients, advocates of patients, and health care providers and administrators. Complementing the call for these changes in institutional structure and power is what I call a *feminist clinical epistemology* (Cutter, 1997), one that offers gender-dependent accounts of clinical reality. A starting point of a feminist clinical epistemology is a focus on questions by and from the perspective of women concerning what is known and how and by whom in order to uncover assumptions about sex and gender and the workings of power and dominant ideologies at play in knowledge claims. In other words, a feminist clinical epistemology may begin to set forth the relations between and among the descriptive, evaluative, and social commitments that frame knowledge of women in clinical medicine. A consequence of this is to critique assumptions and claims that lead to the treatment and mistreatment of women in medicine.

To begin with, one might consider appropriate versus inappropriate kinds of gender bias in medicine. It is one thing to say that lumps on a woman's uterus are likely signs of disease, and another to say that a uterus (or "womb") in *any* woman is cause for disease. For another example, it is one thing to assert that women experience heart disease, and another to say that we can infer that women experience heart disease because men do. The point here is that differences between appropriate and inappropriate gender bias turn on the relevancy of the criteria that are employed in the judgment. Further, the relevancy of the criteria is not simply a matter of accepted epistemological standards but a function of numerous metaphysical, axiological, and cultural assumptions, all of which may require reassessment.

At first, one will need to navigate judgements concerning appropriate and inappropriate bias or discrimination. On the one hand, there are criteria that are inappropriate, or irrelevant, to the situation under study. To take a simple example, deciding who is eligible for a driver's license based on skin color is inappropriate because the criteria used (i.e., skin color) is irrelevant to driving. On the other hand, deciding who is eligible for a driver's license based on ability to see illustrates the use of appropriate criteria that are relevant to the situation. With regard to disease, gender matters insofar as it is a *relevant* criteria in classifying, diagnosing, treating, or researching disease.

As an example, any and all reproductive issues particular to women will necessarily involve issues of gender. How cervical cancer (Chapter 7), for example, is classified, diagnosed, and treated turns on a set of biological criteria regarding cell location and size as well as judgments concerning avoiding false positives and false negatives for purposes of saving lives. It will be important in these and related cases of reproductive disease to assess whether gender is used in an appropriate way at the various stages of the disease. One might explore the extent to which gender enters into assessments about how much women's lives are worth and therefore worthy of protection by public or private programs that offer education, screening, and/or treatment. It will be important as well to evaluate how data on reproductive disease is used in order to prevent the perpetuation of stereotypes.

But women are not solely their reproductive systems; they are not reducible to their uteruses or their X chromosomes (Wolf, 1996). Consider three clinical cases of disease (AIDS, alcoholism, and heart disease) that are not based in the reproductive systems and that lend themselves to further considerations regarding the role gender plays in the construction of disease. The first case, AIDS, illustrates how the *absence* of gender considerations that affect women leads to problems regarding how medicine constructs and implements its clinical nosology. The second case, on the genetics of alcoholism, illustrates how the *inclusion* of gender considerations leads to stereotypes of women that are misleading and harmful. The third case, heart disease, involves what appears to be an appropriate appeal to gender in the construction of disease categories. The challenge before us is to navigate the use of gender considerations in appropriate ways that lead to better medical knowledge about and medical care for women.

4.1. AIDS

The year is 1993. Because medicine initially constructs AIDS as a disease of gay men (Chapter 2), it takes a long time (about 10 years) for clinicians even to acknowledge that the disease can affect women. The symptoms that are used as the basis of diagnosis and qualification for treatment are those that men experience. For many years, symptoms that are unique to women (e.g., pelvic inflammatory disease and cervical cancer) are not defined as AIDS-related. Although a case is identified in a women in 1981, the first year the disease is named, it takes a decade before the distinct patterns of the illness in women are investigated or classified as AIDS and, even then, the disease is still likely to be understood in the West as one that principally affects men. This view proves deadly for many women, since the medical community generally fails to advise women of the risks that they face and fails to identify the signs of illness once infection takes hold. Moreover, clinical interest in tracking the rate of infection among women is largely aimed at keeping an eye on the virus's heterosexual spread and at preventing women from passing the infection onto their children. That is, the focus is on treating women merely as "vectors" of transmission of the virus (Faden and Kass, 1996). Indeed, the more substantial medical and ethical research into the impact of HIV and AIDS on women has been directed at the threat infected women pose to fetuses and infants.

There has been insufficient medical concern about the rate of AIDS among women themselves (Goldsmith, 1992). This is reflected in a relative paucity of data on the course of the disease in women, and a situation where clinical trials are overwhelmingly directed at investigating the effectiveness of various agents on men, not women. Very little information about transmission from woman to woman is available, and virtually no guidance is offered at to what constitutes unsafe sex practices among lesbians. One may, perhaps, welcome women's lack of association with HIV transmission and AIDS, for women have historically been held responsible for the transmission of sexual diseases. Nevertheless, Overall (1995) attributes this systematic neglect of the situation of women infected by HIV to oppressive gender and sexual orientation stereotypes that reflect the dangerous view, endemic to medicine generally, that women and their illnesses are less important than men and their health needs (Sherwin, 2001, pp. 356-360).

The case of AIDS illustrates how appropriate gender considerations fail to operate in the development and implementation of a clinical nosology. From the standpoint of a feminist clinical epistemology, the exclusion of women in medical research and public health initiatives leads to medicine's failure to respond appropriately to AIDS.

4.2. Alcoholism

For a second example, consider how initially appropriate gender considerations lead to problematic gender stereotypes. New knowledge in genetics leads to a host of research on the genetic basis of particular conditions (e.g., colon cancer, schizophrenia, depression, and alcoholism) (Chapter 9). Scientific advances over the past twenty years have shown that drug addiction is a chronic, relapsing disease that results from the prolonged effects of drugs on the brain (Leshner, 1997). This is in contrast to a previous view that holds that addiction is a social or moral problem (Trotter, 1804), which can be handled only with social or moral solutions such as the criminal justice system or moral therapy. Recent work on the genetic basis of alcoholism finds gender differences in the personality traits of alcoholics (Cloninger et al, 1996). More specifically, people dependent on alcohol tend to display the extremes of gender characteristics: aggressiveness and impulsiveness for men, and emotionality and neuroticism for women. Type-I alcoholism is characterized by high levels of anxiety, attentiveness to detail, emotional dependence, an eagerness to help others, rigidity, and sentimentalism (Cloninger et al 1996, pp. 20-21). It is genetically mediated, though not to the same degree as Type II alcoholism, which is "only weakly [influenced] by environmental factors" (Cloninger et al, 1996, p. 18). The latter is an earlier onset form of alcoholism and is correlated with such personality traits as emotional detachment, tough mindedness, impulsiveness, lack of inhibition, and aggressiveness. The phenotype of Type II alcoholism is common only in males, whereas both males and females exhibit Type I alcoholism. Another behavioral trait that has been linked only with Type II alcoholism is sexual promiscuity (Mahowald, 2000, p. 250).

Note that the research on the genetics of alcoholism is that genetics plays a lesser role in female alcoholism than in male alcoholism. Some researchers have challenged that finding and argued for a new type of alcoholism in women: an early onset type that has a strong genetic component (Hill, 1995). Although similar in some ways to Type II alcoholism in men, it does not mimic it entirely. The claim is not that women who have the new type of alcoholism have the personality traits associated with Type II, the claim is that it makes little sense to find such broad genetic difference to addiction among men and women.

By claiming that alcoholism is a disease, that it is genetic, and that its genetics leads to certain phenotypes are stereotypic for men (aggressiveness) and women (neuroticism), genetic research on alcoholism sets up a host of concerns from the standpoint of a feminist clinical epistemology. For female alcoholics, it creates an association between alcoholism and neuroticism and other typical female stereotypes (e.g., high level of anxiety, emotional dependence). For a model that is suppose to minimize social stigmas (e.g., those related to alcoholism being a moral problem), it in fact may encourage social stigmas (e.g., typical gender stereotypes). In addition, it gives the message that genes are deterministic, a claim that has many problems (Chapter 9), especially when applied to women who were for centuries thought to be determined by their reproductive systems. Moreover, it employs terms describing gender traits that are broad, easily misinterpreted, and in the end scientifically-speaking non-falsifiable. In the end, such supposedly "gender-sensitive" research does more harm than good to women.

4.3. Heart Disease

A third example illustrate many of the goods that may arise from including gender in clinical studies. The Women's Health Initiative is a 15-year-health study undertaken by the National Institutes of Health (1993) that includes controlled clinical trials examining strategies to prevent heart and other diseases in postmenopausal women. We have already heard about the absence of women in heart studies and the harms that accrued. Things have changed. So far, study results show that coronary artery disease is the leading cause of death in women. Heart disease kills more women than all cancers combined. Women are twice as likely as men to die within a year after having a heart attack and are also at greater risk of having a second heart attack. Males are more at risk than women at a younger age, but after menopause the incidence of heart disease in men and women is nearly the same. African-American women are sixty percent more likely to die of heart disease than white women (Colorado Hospital Association, 1997). In these studies, traditional risk factors such as family history and cigarette smoking are considered along with others such as hormonal status, diet, exercise, and lifestyle (e.g., whether one is in a care relationship). Recommendations from studies encourage women to listen to their unique signs and symptoms for heart disease, take more time to exercise, relieve stress, and find assistance in their care-taking roles.

Such studies on heart disease in women illustrate how gender can be understood appropriately to include biological forces (e.g., how hormones affect organ function), as well as cultural ones (e.g., women's social roles as caretakers) as well. Unlike studies on alcoholism, variables appear to be definable and testable. ADD Studies such as this are bound to benefit women in terms of education, diagnosis, and treatment.

5. CLOSING

Conceptually speaking, medicine and the clinical classifications that it fashions has been shown to be a non-neutral, or more specifically a non-gender-neutral, endeavor. Practically speaking, the current search to understand and to aid women's disease and health will benefit from a reformulation of the presumptions and approaches embraced by medical research, clinical care, and health care policy. A feminist clinical epistemology is offered as one step in the direction of reformulating medicine for the twenty-first century, which may ultimately lead to a carefully-crafted gendered-dependent approach to disease. It is offered as well as an example of how we may begin to reframe how we understand disease in the twenty-first century.

CHAPTER ENDNOTES

1. This essay was initially developed for the 1997 University of Colorado at Colorado Springs Women's Studies Annual Lecture. I am indebted to Harlow Sheidley, Department of History, Raphael Sassower, Department of Philosophy, and the students in "Women's Health and Disease" (Fall 1998) for their comments and suggestions.

2. Note the rise of women's health centers in major academic institutions and local communities across the U.S. in the late twentieth and early twenty-first centuries.

3. The interplay among gender, ethnicity, and class is complex. Gender is a term used by sociologists to refer to certain social categories: masculinity, femininity, etc. These categories refer to a complex set of characteristics and behaviors prescribed for a particular sex by society (e.g., aggressiveness, nurturing) and learned through the socialization experience. Ethnicity designates sociological divisions of groups of people based on physical

characteristics, language, and customs (e.g., African-American, Irish-American). Class refers to a group of people as a unit according to economic, occupational, or social status (e.g., rich, poor). While it is traditional in modern western thought to associate strength with the masculine, one notes that black slave women in 19th century America were considered strong. Female college professors are considered upper class even though they earn no more than an experienced plumber (Rothenberg, 2001, p. 9). More work is needed on the interplay among gender, ethnicity, and class, especially in the context of disease and health. Interesting work has been done studying anorexia nervosa, which in the DMS is considered a mental disorder. Thompson (1994), for example, provides data to support the view that eating disorders begin as ways women cope with various traumas, including sexual abuse, racism, classism, sexism, heterosexism, and poverty. An understanding of the relation between trauma and the onset of eating disorders will lead to better ways of responding to a condition that affects women across racial and class boundaries.

4. In fact, since at least the time of Hippocrates (approx. 460-377 B.C.) (1943) and Aristotle (384-322 B.C.) (1984), medicine has used gender to justify its actions towards women (Tuana, 1988).

5. Sex (L. *sexus*, sex) is a term used by biologists to refer to certain biological categories: male and female. Identification of sex is base on key factors: chromosomal patterns (e.g., XY, XX), hormonal make-up (e.g., testosterone, estrogen), and genital structure(e.g., penis, clitoris).

6. This distinction can be traced to the Ancient times (Fausto-Sterling, 1987, pp. 62-69), but is not original to medicine.

7. A hermaphrodite possesses some ovarian and some testicular tissue.

8. A male pseudohermaphrodite has testes and some aspects of the female genitalia but lacks ovaries.

9. A female pseudohermaphrodite has ovaries and some aspects of the male genitalis but lacks testes.

10. Here it is interesting to reflect on the implication of cloning from the standpoint of revising our understanding of sexuality and gender, and the power that women have (assuming the technology) in being able to clone without assistance from men.

11. Nevertheless, it is important to recognize the absence of any single gender-norm cross-culturally and particularly within a given culture (Veatch 1992; Spector, 1996; Osborne, 2001).

CHAPTER 11

CONCLUSION

To assess the character of disease, we have looked to a range of philosophical views regarding the nature of and methods of knowing disease. To assess its significance, we explored a range of conceptual accounts regarding the role values play in knowing disease. We found that in our analyses, an account of disease in terms of limited realism, representative realism, and limited stipulative values is unavoidable. This is the case because disease is an evolving construct that provides structure and significance to clinical reality and patient complaints. Disease tells us something about the structure of reality, the ways we know it, the ways we control and manipulate it, and the ways we evaluate it. As a consequence, we come to construe disease as the outcome of the choices among various communities of individuals. The more one recognizes these conditions, the more medicine will enhance human goals. This account sets disease within the broader scope of human endeavors. It is offered as a thoughtful expression of ourselves as knowers and manipulators of clinical reality. It is offered as a basis for further reflection on the ways in which we explain disease to ourselves, to patients and related parties, and to those who influence and are influenced by health care policy.

1. PUTTING IT ALL TOGETHER

In studying disease, one is introduced to the rich domains of medicine and philosophy. Medicine embraces a wide range of endeavors that apply scientific generalizations to the care and cure of problems patients bring to the attention of health care professionals. Medicine refers to the basic medical sciences and theoretical endeavors tied to treatment. One might think here of theories about the ways in which the pancreases works and explanatory models of the development of diabetes mellitus, which are tied to regimens for the control of diabetes mellitus through the use of insulin. As a result, one may have philosophical puzzles about theories of function and disease, about theories of treatment, and the ways in which health care practitioners engage in their preventive and therapeutic activities. Broadly speaking, medicine involves not only what doctors of medicine do, but also the intellectual and practical endeavors of those who participate in medicine.

There are good reasons for viewing this collage of endeavors as clustered around therapy: as enterprises of therapy, enterprises in the prevention of the need for therapy, studied of the biological and/or psychological bases for the occasions for therapy or studies of the efficacy of therapy. "Therapy" has a wide-range of meaning, as does *therapeia* in the Greek, which has a broad scope including a waiting on, a

149

service, an attendance, the service of divine worship, as well as a fostering, nurturing, or tending to in sickness. So also therapy in English includes the senses of tending to the sick as well as treating medically with the intent to cure, heal, or give care. This ambiguity in the word "therapy" is of particular importance, for a great proportion of the "therapeutic" activities of clinicians involve attending to the worried-well and performing non-curative activities, such as prescribing contraceptives, delivering babies, and controlling anxieties through drugs. Therapy means much more than curing and healing.

Therapy is what ties medicine to the clinic, as oppose to the laboratory. By "clinic," I mean the setting in which the medical professional directly encounters the patients for purposes of assessing and manipulating clinical problems. The term "clinical problems" identifies generally those difficulties that stand out as conditions that ought to be addressed and solved by medicine. These difficulties include, among others, disease, illness, deformity, and dysfunction. They involve physiological processes considered as causally determined, not directly or immediately under voluntary control, and recognized as the substrate of dysfunctions, pains, or disfigurements. Disease, as a clinical problem, is an explanatory notion in that accounts of disease set out to make intelligible the pathological findings or processes associated with the complaints or actual mal-experiences of patients. On this view, a critical examination of disease provides insight into the ways in which we explain in medicine.

The dynamic relation between knowing and treating may also be one between knowing and valuing. Some facts in medicine are more important than others not because they help us know more truly but because they help us treat more effectively, where effectiveness is measured in monetary and non-monetary terms. Put another way, clinicians are not so much concerned about whether classifications are true or false but whether they are useful. Under such circumstances, true knowledge is in the service of effective and efficient action. Given this, what it means to know and to intervene effectively will likely vary among communities of clinicians. There are no such thing as neutral, naked, or bare facts. Facts always appear interpreted within the embrace of evaluative frameworks, whether or not these frameworks are formally or informally developed as scientific accounts.

In addition, such frameworks are given in a socio-historical context. There is no placeless and timeless account of reality, or at least there is no such interpretation available for humans. As the history of medicine illustrates, explanations of disease have been changed and altered over time. Sauvages (1768) and Cullen (1769) explained disease in terms of a constellation of signs and symptoms so that they could be identified in the future. These classifications were primarily syndrome-based rather than etiological (causal) or anatomical (appeal to an underlying matrix). With nineteenth century developments in anatomy, pathology, and microbiology, there occurs a tradition in the ways in which disease is explained. Fevers, pain, and diarrhea are no longer considered diseases in their own right, but symptoms associated with an underlying pathoanatomical and pathophysiological basis or process. Clinical complaints that have not previously associated could now to brought together under one rubric.   These changes lead to the notion that the goal of medicine is comprehending pathoanatomical and pathophysiological underpinnings of disease while merely reporting the symptoms is inadequate.

In talking about disease, then, one must identify the particular community of scientists and clinicians to which one wishes to maker reference. Each community is defined by its own rule of evidence, inference, and negotiation. Scientists and clinicians who are members of a particular community know which facts are important and why.   They know when particular facts warrant particular conclusions or particular interventions. In medicine, the rules of evidence will include recipes for action, which incorporate implicit, and at times explicit, value judgements about proper trade-offs between costs and benefits. The character of proper trade-offs will often be negotiated without claiming that correct answers can be discovered. Disease is in part a negotiated reality.

This conclusion does not doom us to a hopeless relativity with regard to claims regarding disease. Facts do share, challenge, and even disconfirm accounts regarding disease. One is compelled to analyze carefully the different roles played by considerations that tend to be more factual (i.e., more oriented around epistemic claims) and those more oriented around intervention (more oriented toward non-epistemic claims). Such is the tedious but important work of scientists and clinicians as they follow the clinico-scientific method in their investigations. We neither know diseases truly nor completely make them up.

All of this, in addition to the previous chapters, is to underscore that a philosophical approach to medical concepts such as disease is unavoidable. The more one comes to understand the conceptual underpinnings of our ideas and actions in medicine, the more one can make sense out of clinical reality and the more one can successfully manipulate clinical reality. We are reminded once again that medicine as a learned profession has much to gain from reflections in the humanities. Along with law and medicine, health care professions have traditionally not been seen merely as technicians or tradesmen, but rather as professionals with special commitments to patients as well as society generally. Learned professions, after all, are those that give an account of themselves and place themselves within the general concerns of human culture.

2. IMPLICATIONS FOR MEDICINE

This account of disease has implications for how we are to view medicine in the twenty-first century. In particular, it leads us to reflect on the grounds and the limits of the medical profession's legitimate authority and its obligation toward patients through the analysis of the role of values in medicine (Cutter, 1993).

The lesson follows. Clinical science employs explanation to render the word intelligible.   Intelligibility, minimally speaking, is tied to intersubjectivity and agreement among cognitive agents.   Intersubjectivity grounds the possibility in medicine of usual and customary standards of care and formally articulated indications for treatment. Agreement makes possible action, which in medicine involves the caring and curing of human disease and illness. In the history and development of modern Western medicine, explanation has a two-tiered character, that of (1) the *clinical* understanding of disease as a syndrome, and (2) the *laboratory* or basic science account, which interprets the clinical elements of a syndrome in terms of a pathoanatomical and pathophysiological correlates with unacceptable pain, disability, dysfunction, or death. That is, with the developments in seventeenth and eighteenth century medicine, accounts of the constellation of signs and symptoms provides the basis of clinicians' understanding of disease.

       With the advent of nineteenth century pathological construals of human illness, clinical signs and symptoms are reinterpreted in terms of the basic laboratory sciences, which have in turn delivered new clinical descriptions, and which are themselves submitted to further revision and given further changes in basic scientific appreciations of disease. Fevers, pain, and fluxes are no longer considered diseases in their own right, but symptoms associated with underlying pathoanatomical, pathophysiological, or pathopsychological processes. Etiological accounts of the origin of the pathoanatomical and pathophysiological findings emerge and have their impact on distinguishing between and among disease conditions. The laboratory or basic scientific account of disease provides, in short, theoretical bases through which to organize and reorganize descriptive accounts of patient complaints in terms of underlying pathoanatomical, pathophysiological, and pathopsychological correlates.

       The interaction between the two levels is dynamic. Clinical facts and scientific theories of disease interplay in such a way as to require multi-factorial accounts of clinical findings. On the one hand, the basic scientific level tends to interpret the world of clinical complaints in terms of a number of pathoanatomical, pathophysiological, and pathopsychological correlates so that often no single factor stands out as both necessary and sufficient. On the other hand, the clinical level tells the laboratory scientist what is pathological and thus brings direction to medical problems. This two-tiered, dynamic account of clinical problems provides, in short, ways in which to organize and predict clinical reality.

       On this analysis, clinical problems are not simply discovered but are in part the product of the creative enterprise we call medicine. In medicine, clinical reality is constituted for practitioners in the sense that the world is seen through prevailing theoretical frameworks. One can never know truly, undistorted by particular frameworks. As Kant reminds us, "What the things-in-themselves may be I do not know, not do I need to know, since a thing can never come before me expect in experience" (Kant, 1929 [1787], A277-B333). Clinicians' observations and goals guide the construction of disease nosologies and nosographies, which change with time. Clinical reality is created in the sense that it is seen and interpreted within the embrace of theoretical frameworks.

       Yet, all is not constructed. The world of bodily dysfunction and patient complaints is not a figment of individual imaginations. Human beings, by virtue of their shared biological and psychological nature and evolutionary development, experience similar kinds of physical and mental challenges, much of which have become the focus of scientific and medical research and practice. Clinical reality is discovered in the sense that it is given within the embrace of theoretical frameworks.

Insofar as medicine is tied to action for the sake of treating patient complaints and achieving goods, instrumental values direct clinical explanation. Clinicians explain clinical problems in terms of certain goals, e.g., maximizing benefits and minimizing burdens (e.g., costs and harms) to particular patients. Such consequential or outcome language underlies the enterprise of manipulation or intervention in medicine. Intervention–or action--in medicine is devoted to the achievement of goals through means that are developed for effectively and efficiently obtaining the goals. Note the dynamic character of this dimension. It interprets models and modes of examining medical phenomena in terms of therapeutic goals, such as providing more encompassing explanations of disease phenomena and in the end more useful forms of treatment. Seventeenth and eighteenth century medicine delivered, at best, symptomatic treatments (e.g., herbs, purges, cold baths) for a clinical problem. The clinical problem (e.g., pains) was that which was to be manipulated and was seen in the context of the patient's whole person.

With developments in nineteenth and twentieth century medicine, a critical dialectic is initiated between the world of symptomatic treatment and that of etiologic intervention. Medicine no longer simply specifies that which was to be manipulated, the basis for medicine intervention. Rather, medicine offers various accounts of that which manipulates, treatments that are supported by theory, by etiologic interpretation. An interplay between clinical practitioners and biomedical scientists emerges in which the basic sciences provide the bases for realizing certain ends (i.e., the treatment or prevention of patient complaints) by structuring the character of interventions, of that which manipulates.

Since all clinicians and patients do not share the same views regarding which clinical goals are to be achieved and why, and since cooperation among physicians, patients, and third parties requires agreement concerning understandings about what are "good" goals, non-moral and moral values play a role in the formulation of clinical explanation and manipulation. These values are varied and complex, and include instrumental (which tell us how to get from means to end), aesthetic (which tell us what ideals of form and grace are worthy of achievement), functional (which tell us what ideals of activity are proper to an organism), and ethical (which tell us what ought to be done and by who). Treatment in medicine emphasizes the central role values play in our understanding of clinical reality. It emphasizes the practical and moral grounds and limits of the medical profession's obligation toward patients.

Explanation and intervention–theory and practice, science and technology, diagnosis and therapeutics–are uniquely interwoven in medicine. Medicine provides the forum for this interaction as it attempts to know more fully and intervene more carefully for purposes of alleviating human pain, dysfunction, and inevitably death.

3. THE NEED FOR PHILOSOPHY OF MEDICINE

This project on disease is part of a greater field called philosophy of medicine. Scholarship in philosophy of medicine indicates an internationally active, vital field, and documents the increasing importance of philosophical reflection for all dimensions of science, practice, and policy in the health care arena. Despite the evident interest in philosophy of medicine, there are two concerns that drive the need for this work on disease. First, there has been in the last three decades, an explosion of interest in one of philosophy of medicine's sub-specialities, bioethics. This explosion has directed interest toward numerous applied issues in health care at the cost of the conceptual issues addressed in philosophy of medicine. This is the case in part because of the contemporary appeal for quick, practical answers to difficult questions and because conceptual presuppositions have been assumed away. Only in cases of crisis do we reconsider them. A refocus of attention on issues in the philosophy of medicine, such as disease, is necessary in order to redirect interest to the conceptual issues informing the practical ones, because it is our reliance on these issues that dictates our clinical and policy applications.

Second, and in part due to insufficient attention, there are numerous unsettled issues in philosophy of medicine. These include, but are not limited to, the nature and scope of philosophy of medicine, the character of medical theory and practice, the extent to which health and disease are definable, and the likely prospects of medicine in the Third Millennium, especially given the rise of molecular biology. It is thus important to have a sense of the field of philosophy of medicine at large as a broader framework, even if the interest is to focus on select elements, such as the nature of disease, in order to reconstitute the field.

In order to respond to these two-fold concerns (lack of attention with dire practical consequences and loss of focus in bioethics on critical conceptual issues), while simultaneously celebrating the richness and diversity of the field in general, this work takes stock of disease in the broadest sense of the term from a philosophical perspective. A central conviction that informs this project is that the older paradigm, when humanities and philosophical reflection were regarded as external and somewhat secondary to medical science and practice, can no longer be sustained. Rather, concerns in medical science and practice can no longer be seen as separate and distinct from concerns raised in humanities. Philosophy of medicine offers the opportunity, through an interdisciplinary approach, to link the concerns and develop the discussions.

An additional insight reemerges. Despite the hard work of Pellegrino (1998), Engelhardt (1996), Veatch (1995), McCullough (1998), and Khushf (1997), few scholars appreciate that exploration of bioethical issues requires a consideration of the actual transactions and conditions of medical practice, the historical development of medical concepts and practices, and the ontological and epistemological controversies that influence the ethical and public policy debates in medicine. Such are the concerns of philosophy of medicine and if left unpursued will leave bioethics and other areas such as health policy bankrupt in its ability to move discussions beyond its particular utilitarian focus.

4. THE NEED FOR ANALYSES OF DISEASE

As this work illustrates, disease is central to clinical medicine. In clinical medicine, disease brings intelligibility to a wide range of clinical problems. Put another way, disease is an explanatory notion in that accounts of disease set out to make intelligible the pathological findings and processes associated with the complaints or actual mal-experiences of patients.

The ways in which disease brings sense to clinical reality may be resolved into two major human endeavors, knowing and valuing. Clinicians wish to establish though disease the regularities of occurrences among clinical phenomena and to find simple and coherent models to account for these regularities. In this way, clinicians function as medical scientists seeking to know the facts that constitute the world.

Our consideration of the metaphysical dimension of disease focuses generally on theory structure and change during the development of modern medicine. This framework can be used to investigate further the character of clinical theory. One might take a closer look at what constitutes progress and development in particular areas of medicine, such as genetics and immunology. In doing so, the history of medicine must not be forgotten. As Lakatos (1970) reminds us, history without philosophy lacks form; philosophy without history lacks content. Put another way, insofar as medicine is a practical endeavor, devoted to the care and cure of patient vexations, concepts in medicine are pragmatic and rooted in the history and traditions of clinical practice.

Our study of the epistemological character of disease focuses on the relation between the knower and known in understanding disease. Issues regarding theory confirmation and falsification, what constitutes evidence and proper rules for inference and causal claims, and the central role of statistical explanation in medicine are topics that deserve greater attention. Further work is needed as well on the structure of medical decision-making (Pellegrino, 1979; Laor and Agassi, 1990), a topic that is of importance particularly to those interested in the logic of medicine and its computerization. However, ingredient to this task of explaining medical reality are judgments regarding what goods patients seek, what general accounts show the rationality of such goods, and what constitutes appropriate means to achieve such goods. Such judgments are those of good and bad, and right and wrong, or praise, in that agents capable of choice select clinical goods.

Our study of the axiological character of disease focuses, then, on the role values play in our understanding and manipulation of disease. This framework may be used to study in greater detail the meaning of *praxis* and *poiesis*, doing and making, or action and production, in the clinical setting (Wartofsky, 1975, 1976). Put another way, medicine offers an opportunity to study the relation between theory and practice in Western thought. No doubt, there will be challenges. The boundaries and limits of good and prudent actions in medicine become particularly troublesome in light of the rise of high-technology medicine (Reiser, 1978), the contemporary transient nature of the patient-clinician-health care institution relationship, and the tensions between medical practice and medical economics (Morreim, 1997). The ways in which clinicians weigh transaction and opportunity costs in the treatment of particular diseases turn on concerns to minimize costs (e.g., morbidity, mortality, financial) to patients and maximize outcome (e.g., opportunities to pursue individual wishes, desires, and goals) for purposes of respecting persons. The bioethical challenges are indeed daunting in medicine in the Third Millennium.

In short, clinical explanation involves a complex interplay between two major intellectual goals, namely, 1) to know the fabric of clinical reality, and 2) to know the goods of human life particularly as they have to do with the mitigation, if not elimination, of patient complaints. The dynamic interactions between theory and practice, and knowing and valuing, may be resolved further. Some facts are more important than others, not because they help us to know more, but because they help us to treat more effectively, where effectiveness is measured in monetary and non-monetary terms. As Wulff (1981b) has argued, clinicians are not so much concerned about whether classifications are true or false, but whether they are useful. Physicians tend to be pleased with an artificial, instrumental classification even if it strategically distorts reality, as long as it maximizes the effectiveness of therapeutic interventions. In the applied sciences, such as medicine, epistemic, or knowledge-gathering, goals are important, but non-epistemic, or action-oriented, goals are central. The point is that facts in medicine are always interpreted within human contexts involving goals to be achieved.

Taken together, the metaphysical, epistemological, and axiological dimensions of disease lead to a contextual account. A contextual account of disease raises additional concerns tied to the extent to which socio-cultural influences condition our knowledge and actions in medicine. One might wish to take a closer look at the relation between notions of causation and causal responsibility in medicine. Given that we in part create clinical, causal explanations, and given that we are in part responsible for our creations, what can be said about the ways in which we assign causal accountability in medicine? What is the relation, if any, between the ways in which we explain disease and the ways in which we assign blame and praise to individuals? Notions of causation in medicine may indeed shed light on our understanding of clinical morality and the role of values in clinical diagnosis. The extent to which public policy, socio-cultural, political, economic, ethnic, class, and gender influences condition disease are in need of greater discussion. In the onset of a global economy and increased demographic shifts in the world, for example, work on disease can modestly serve as a tool for discussions across borders regarding how we understand and undertake the human condition in the medical arena.

The foregoing issues are suggested as directions for future inquiry. These were not attempted in this study in any detail because of the limited scope of this work, which set out to display generally the major conceptual features of disease. The framework offered here lays a basis for future work on issues in the philosophy of medicine, and its offspring, biomedical ethics, and on the character of disease and that which is affected by its designation.

REFERENCES

Achinstein, Peter (1983). *The Nature of Explanation.* New York: Oxford University Press.

Achterberg, Jeanne (1991). *Woman as Healer: A Panoramic Survey of the Healing Activities of Women From Prehistoric Times to the Present.* Massachusetts: Shambhala Press.

Agich, George (1999). The Importance of Management for Understanding Managed Care. *The Journal of Medicine and Philosophy, 24, 5,* 518-534.

Akerknecht, Edwin A. (1982). *A Short History of Medicine.* Baltimore, Maryland: Johns Hopkins University Press.

Albert, Daniel A. et al. (1988). *Reasoning in Medicine: An Introduction to Clinical Inference.* Baltimore, Maryland: Johns Hopkins University Press.

American Association for the Advancement of Science (1989). *Science For All Americans.* Washington, D.C.: American Association for the Advancement of Science.

American Psychiatric Association (1952). *Diagnostic and Statistical Manual of Mental Disorders.* Washington, D.C.: American Psychiatric Association.

American Psychiatric Association (1968). *Diagnostic and Statistical Manual of Mental Disorders.* 2nd ed. Washington, D.C.: American Psychiatric Association.

American Psychiatric Association (1980). *Diagnostic and Statistical Manual of Mental Disorders.* 3rd ed. Washington, D.C.: American Psychiatric Association.

American Psychiatric Association (1987). *Diagnostic and Statistical Manual of Mental Disorders.* 3rd ed. rev. Washington, D.C.: American Psychiatric Association.

American Psychiatric Association (1994). *Diagnostic and Statistical Manual of Mental Disorders.* 4th ed. Washington, D.C.: American Psychiatric Association.

American Psychiatric Association (2000). *Diagnostic and Statistical Manual of Mental Disorders.* 4th ed. rev. Washington, D.C.: American Psychiatric Association.

Antlers, Joyce and Fox, Daniel M. (1978) . The Movement Toward a Safe Maternity: Physician Accountability in New York City, 1915-1040. In: J.W. Leavitt and R.L. Numbers (Eds.), *Sickness and Health in America: Readings in the History of Medicine and Public Health* (pp. 375-392). Wisconsin: University of Wisconsin Press.

Aquinas, St. Thomas (1994). *Summa Contra Gentiles.* In: F.S. Baird and W. Kaufman (Eds.), *Medieval Philosophy* (pp. 36-49). New Jersey: Prentice-Hall Publishing Company.

Aquinas, St. Thomas (1994). *Summa Theologica.* In: F.S. Baird and W. Kaufman (Eds.), *Medieval Philosophy* (pp. 50-61). New Jersey: Prentice-Hall Publishing Company.

Aristole (1941). *On the Soul.* In: R. McKeon (Ed.), *The Basic Works of Aristotle* (pp. 535-603). New York: Random House.

Aristotle (1975). *Posterior Analytics.* Trans. J. Barnes. Oxford: Clarendon Press.

Aristotle (1984). *On the Generation of Animals*. In: J. Barnes (Ed.), t rans. A. Platt, *The Complete Works of Aristotle*, Vol. 8 (pp. 665-683). New Jersey: Princeton University Press.

Aristotle (1985). *Nicomachean Ethics*. Trans. T. Irwin. Indiana: Hackett Publishing Co.

Aronowitz, Robert (1998). *Making Sense of Illness: Science, Society, and Disease*. London: Cambridge University Press.

Audi, Robert (Ed.) (1995). *The Cambridge Dictionary of Philosophy*. England: Cambridge University Press.

Automobile Workers v. Johnson Controls 111 S. Ct. 1996, 1991.

Ayer, Alfred J. (1935). *Language, Truth, and Logic*. Cambridge: Oxford University Press.

Bacon, Francis (1989 [1620])) *The New Organon*. In: M.R. Matthews (Ed.), *The Scientific Background to Modern Philosophy* (pp. 45-52). Indiana: Hackett Publishing Co.

Baltimore, D. and Heilman, C. (1998). HIV Vaccines: Prospects and Challenges. *Scientific American, 279, 1*, 98-103.

Barondess, J.A. (1979). Disease and Illness--A Crucial Distinction. *American Journal of Medicine, 66*, 375-386.

Barre-Sinoussi, F. et al. (1983). Isolation of T-lymphotropic Retrovirus From a Patient at Risk for Acquired Immune Deficiency Syndrome (AIDS). *Science, 220*, 868-871.

Bartholomew, Robert E. (2000). *Exotic Deviance: Medicalizing Cultural Idioms*. Colorado: University of Colorado Press.

Bartlett, J.G. and Moore, R.D. (1998). Improving HIV Therapy. *Scientific American, 279, 1*, 84-93.

Baumgarten, M. (1954). *Reflections on Poetry*. Trans. K. Aschenbrenner and W.B. Holt. California: University of California Press.

Bayer, Ronald (1981). *Homosexuality and American Psychiatry: The Politics of Diagnosis*. New York: Basic Books.

Beauchamp, Tom L. (Ed.) (1974). *Philosophical Problems of Causation*. California: Dickenson Publishing Co.

Beauchamp, Tom L. (1991). *Philosophical Ethics: An Introduction to Moral Philosophy*. 2nd ed. New York: McGraw-Hill Company.

Beauchamp, Tom L. (1987). Ethical Theory and the Problem of Closure. In: H.T. Engelhardt, Jr. and A.L. Caplan (Eds.). *Scientific Controversies: Case Studies in the Resolution and Closure of Disputes in Science and Technology* (pp. 27-48). Massachusetts: Cambridge University Press.

Beauchamp, Tom L. and Childress, James F. (2001). *Principles of Biomedical Ethics*. 5th ed. New York: Oxford University Press.

Beauchamp, Tom L. and Rosenberg, Alexander (1981). *Hume and the Problem of Causation*. New York: Oxford University Press.

Beckwith, Jon and Alper, Joseph S. (1998). Reconsidering Genetic Antidiscrimination Legislation . *Journal of Law, Medicine, and Ethics, 26*, 205-210.

Benedict, Ruth (1934). Anthropology and the Abnormal. *Journal of General Psychology*, 10, 72-74.

Benedict, Ruth (1934). *Patterns of Culture*. New York: Houghton Mifflin.

Bennett, William (Ed.) (1993). *The Book of Virtues: A Treasury of Great Moral Stories*. New York: Simon and Schuster.

Bentham, Jeremy (1948 [1789]). *Introduction to the Principles of Morals and Legislation*. Ed. J. LaFleur. New York: Hafner.

Bergler, E. (1956). *Homosexuality: Disease or Way of Life*. New York: Hill and Wand.

Bergson, Henri (1944). *Creative Evolution*. Trans. A. Mitchell. New York: The Modern Library.

Berkeley, George (1974 [1710]). *A Treatise Concerning the Principles of Human Knowledge*. In *The Empiricist*, pp. 135-215. New York: Doubleday.

Bernard, Claude (1957 [1865]). *An Introduction to the Study of Experimental Medicine*. New York: Dover Publications, Inc.

Bernstein, Richard J. (1971). *Praxis and Action*. Pennsylvania: University of Pennsylvania Press.

Berwick, Donald M. and Weinstein, Milton C. (1985). What Do Patients Value?: Willingness to Pay for Ultrasound in Normal Pregnancy. *Medical Care*, 23, 7, 881-893.

Bichat, Xavier (1801). *Anatomie gènèral appliquèe à la physiologie et la mèdicine*. 4 Vols. Paris: Brosson, Gabon.

Bichat, Xavier (1981 [1827]). *Pathological Anatomy: Preliminary Discourse*. In: A.L. Caplan et al. (Eds.), *Concepts of Health and Disease: Interdisciplinary Perspectives* (pp. 167-173). Massachusetts: Addison-Wesley Publishing Company.

Bickel, Janet (1986). *Integrating Human Values Teaching Programs Into Medical Students' Clinical Education*. Washington, D.C.: Association of American Medical Colleges.

Billings, Paul et al. (1992). Discrimination as a Consequence of Genetic Testing. *American Journal of Human Genetics*, 50, 276-282.

Biological Sciences Curriculum Study (1992). *Mapping and Sequencing the Human Genome: Science, Ethics, and Public Policy*. Colorado: Biological Sciences Curriculum Study.

Biological Sciences Curriculum Study (1996). *The Human Genome Project: Biology, Computers, and Privacy*. Colorado: Biological Sciences Curriculum Study.

Biological Sciences Curriculum Study (1997). *The Puzzle of Inheritance: Genetics and the Methods of Science*. Colorado: Biological Sciences Curriculum Study.

Biological Sciences Curriculum Study (2000). *Genes, Environment, and Human Behavior*. Colorado: Biological Sciences Curriculum Study.

Biological Sciences Curriculum Study (2001). *The Brain: Understanding Neurobiology Through the Study of Addiction*. Colorado: Biological Sciences Curriculum Study.

Biological Sciences Curriculum Study (2002). *Bioinformatics and the Human Genome Project*. Colorado: Biological Sciences Curriculum Study.

Black, Henry Campbell (1979). *Black's Law Dictionary*. 5th ed. Minnesota: West Publishing Company.

Blaxter, Mildred (1983). The Causes of Disease: Women Talking. *Social Science and Medicine, 17, 2,* 59-69.

Boerhaave, Hermann (1742-1746) (1708). *Institutes, Dr. Boerhaave's Academical Lectures on the Theory of Physics*. 6 Vols. London: W. Innys.

Bole, Thomas J. (1995). The Neologism *Ontoi* in Broussais's Condemnation of Medical Ontology. *Journal of Medicine and Philosophy, 20, 5,* 543-549.

Bondeson, William and Jones, James (Eds.) (2002). *The Ethics of Managed Care: Professional Integrity and Patient Rights*. Holland: Kluwer Academic Publishers.

Boorse, Christopher (1975). On the Distinction Between Health and Disease. *Philosophy and Public Affairs, 5,* 49-68.

Boorse, Christopher (1977). Health as a Theoretical Concept. *Philosophy of Science, 44,* 542-573.

Borysenko, Joan (1996). *A Woman's Book of Life: The Biology, Psychology, and Spirituality of the Feminine Life Cycle*. New York: Riverhead Books.

Boston Women's Collective (1973). *Our Bodies, Ourselves*. New York: Simon and Schuster.

Boston Women's Collective (1998). *Our Bodies, Ourselves for the New Century*. 3rd ed. New York: Simon and Schuster.

Bowman, Inci (1976a). Classification of Diseases.: Part I. *The Bookman, 3, 6,* 1-10.

Bowman, Inci (1976b). Classification of Diseases: Part II. *The Bookman, 3, 7,* 1-12.

Bowman, Inci (1976c). Classification of Diseases, Part III. *The Bookman, 3, 8,* 1-10.

Brandt, J. et al. (1989). Presymptomatic Diagnosis of Delayed Onset Diseases with Linked DNA Markers: The Experience of HD. *Journal of the American Medical Association, 261,* 3108-3114.

Brody, Baruch A. (1988). *Life and Death Decision Making*. New York: Oxford.

Brooks, C.M. and Cranefield, P.F. (Eds.) (1959). *The Historical Development of Physiological Thought*. New York: Hafner.

Broussais, F.J.V.: (1821). *Examen des doctrines médicales et des systéms de nosologie*. 2 Vols., Paris: Méuignon-Marvis.

Broussais, F.J.V. (1981 [1828]). *On Irritation and Insanity*. In A.L. Caplan et al., *Concepts of Health and Disease: Interdisciplinary Perspectives* (pp. 355-360). Massachusetts: Addison-Wesley Publishing Company.

Broverman, I.K. et al. (1970). Sex-Role Stereotypes and Clinical Judgments of Mental Health. *Journal of Consulting and Clinical Psychology,* 34, 1-7.

Brown, John (1803). *The Elements of Medicine*. Translated from the Latin, with comments and illustrations, by the author. New Hampshire: William and Daniel Treadwell.

Brunner, H.D. et al. (1993). X-Linked Borderline Mental Retardation With Prominent Behavioral Disturbance: Phenotype, Genetic Localization, and Evidence for Disturbed Monoamine Metabolism. *American Journal of Human Genetics*, *52*, 1032-1039.

Buchanan, Scott (1938). *The Doctrine of Signatures*. London: K. Paul.

Callahan, Daniel (1999). Medicine and the Market: A Research Agenda. *The Journal of Medicine and Philosophy*, *24, 3,* 224-242.

Callahan, Joan C. (1988). *Ethical Issues in Professional Life.* New York: Oxford University Press.

Cambien, F. et al. (1992). Deletion Polymorphism in the Gene for Angiotensin Converting Enzyme a Potent Risk Factor for Myocardial Infarction. *Nature, 359,* 5641-5644.

Cameron, Nigel M. et al.: 2000, *Bioethics and the Post-Christian Society: Moral Engagement and the End of Consensus.* Michigan: Wm. B. Eerdmans.

Campbell, E.J.M. et al.: 1979, The Concepts of Disease. *British Medical Journal, 2,* 757-762.

Camus, Albert (1991). *The Plague.* Trans. S. Gilbert. New York: A.A. Knopf.

Canguilhem, Georges (1978 [1966]). *On the Normal and the Pathological.* Trans. C.R. Fawcett. Holland: D. Reidel Publishing Company.

Caplan, Arthur L. (1992). Does the Philosophy of Medicine Exist? *Theoretical Medicine, 13, 1,* 67-77.

Caplan, Arthur L. (1993). The Concepts of Health, Illness, and Disease. In: W.F. Bynum and R. Porter (Eds.). *Companion Encyclopedia of the History of Medicine.* Vol. I. (pp. 233-248). London: Routledge.

Caplan, Arthur L. (1997). The Concepts of Health, Illness, and Disease. In: R.M. Veatch (Ed.), *Medical Ethics* (pp. 57-73). Boston: Jones and Bartlett Publishers.

Caplan, Arthur L., et al. (Eds.) (1981). *Concepts of Health and Disease: Interdisciplinary Perspectives.* Massachusetts: Addison-Wesley Publishing Company.

Carlson, E.A. (1991). Defining the Gene: An Evolving Concept. *American Journal of Human Genetics, 49,* 475-487.

Carr, Edward H. (1961). *What is History?* New York: Knopf, Chs. II and IV.

Cartwright, Samuel (1981 [1851]).Report on the Diseases and Physical Peculiarities of the Negro Race. In : A.L. Caplan et al.,(Eds.), *Concepts of Health and Disease: Interdisciplinary Perspectives* (pp. 305-325). Massachusetts: Addison-Wesley Publishing Company.

Cavelli-Sforza, L.L. (1986). Cultural Evolution. *American Zoologist, 26,* 845-855.

Cavelli-Sforza, L.L. and Feldman, M.W. (1981). *Cultural Transmission and Evolution.* New Jersey: Princeton University Press.

Ceccarelli,Leah (1995).A Rhetoric of Interdisciplinary Scientific Discourse:Textual Criticism of Dobzhansky's *Genetics and the Origin of Species. Social Epistemology, 9, 2,* 91-111.

Centers for Disease Control (1981a). Kaposi's Sarcoma and Pneumocystic Pneumonia Among Homosexual Men--New York City and California. *Morbidity and Mortality Weekly Reports, 30,* 305-308.

Centers for Disease Control (1981b) Pneumocystic Pneumonia--Los Angeles. *Morbidity and Mortality Weekly Reports, 30*, 250-252.

Centers for Disease Control (1982). Update on Acquired Immune Deficiency Syndrome (AIDS)--United States. *Morbidity and Mortality Weekly Report, 31*, 507-514.

Centers for Disease Control (1983). Immunodeficiency Among Female Sexual Partners of Males with Acquired Immune Deficiency Syndrome (AIDS). *Morbidity and Mortality Weekly Report, 31*, 697-698.

Centers for Disease Control (1985). Revision of the Case Definition of Acquired Immunodeficiency Syndrome for National Reporting--United States. *Morbidity and Mortality Weekly Report, 34*, 373-375.

Centers for Disease Control (1986). Classification System for Human T-Lymphotropic Virus Type III/Lymphadenopathy-Associated Virus Infections. *Journal of the American Medical Association, 256*, 20-25.

Centers for Disease Control (1987). Revision of the Surveillance Case Definition of Acquired Immunodeficiency Syndrome. *Morbidity and Mortality Weekly Report, 36*, 3S-15S.

Centers for Disease Control (1992). 1993 Revised Classification System for HIV Infection and Expanded Surveillance Case Definition for AIDS Among Adolescents and Adults. *Morbidity and Morality Weekly Report, 41*, 1-19.

Centers for Disease Control and U.S. Public Health Services (1998). *What About AIDS Testing?* (Pamplet), Georgia: Centers for Disease Control.

Childs, Barton (1994). A Logic of Disease. In : C. Scriver et al. (Eds.), *The Metabolic and Molecular Bases of Disease*, 7ᵗʰ ed. New York: McGraw-Hill.

Childs, Barton (1999). *Genetic Medicine: A Logic of Disease*. Maryland: Johns Hopkins University Press.

Childs, Barton and Scriver, C.R. (1986) Age at Onset and Causes of Disease. *Perspectives in Biology and Medicine, 29, 3*, 437-460.

Chomsky, Noam (1965). *Aspects of the Theory of Syntax*. Massachusetts: Massachusetts Institute of Technology Press.

Chopra, Deepak (1998). *Healing the Heart: A Spiritual Approach To Reversing Coronary Heart Disease*. New York: Random House.

Cloninger, C. et al. (1996). Type I and Type II Alcoholism: An Update. *Alcohol Health and Resaerch World, 20, 1*, 18-23.

Clouser, K. Danner and Hufford, David J. (Eds.) (1993). Nonorthodox Medical Systems: Their Epistemological Claims. *The Journal of Medicine and Philosophy, 18, 2*, 101-231.

Coffin, J. et al. (1986). Human Immunodeficiency Viruses. *Science , 232*, 308-309.

Cohen, Henry (1981 [1961]). The Evolution of the Concept of Disease. In: A.L. Caplan et al. (Ed.). *Concepts of Health and Disease: Interdisciplinary Perspectives* (pp. 209-219). Massachusetts: Addison-Wesley Publishing Company.

Cohen, Jon (1994). The Duesberg Phenomenon. *Science*, 266, 1642-1644.

Collingwood, R.C. (1940). *An Essay on Metaphysics*. New York: Oxford University Press.

Colorado Hospital Association (1997). *Issue: The Silent Epidemic: Women and Cardiovascular Disease, 4, 2,* 1-2.

Colorado Revised Statute 10-3-1104.7 (On Genetic Testing)

Comarow, Avery (2000). How Bad Science Can Be Hazardous To Your Health. *U.S. News and World Report, May 8,* p. 68.

Complementary and Alternative Medicine (1997). *Park Ridge Center Bulletin, 2* (November/December).

Comte, Auguste (1988 [1830-42]). *Introduction to Positive Philosophy*. Trans. F. Ferre. Indiana: Hackett Publishing Company.

Conner, E.M. et al (1994). Reduction of Maternal and Infant Transmission of Human Immunodeficiency Virus Type I with Zidovudine Treatment. *New England Journal of Medicine, 331, 18,* 1173-1180.

Conte, J.E. (1983), Infection-Control Guidelines for Patients with Acquired Immunodeficiency Syndrome. *New England Journal of Medicine, 309,* 740-744.

Coors, Marilyn (2002). Therapeutic Cloning: From Consequences to Contradiction. *The Journal of Medicine and Philosophy, 27, 3,* 297-317.

Cowan, David et al. (1992). *Teaching the Skills of Conflict Resolution.* California: Interchoice Publishing.

Cowley, Geoffrey, with Hager, Mary and Joseph, Nadine (1990). Chronic Disease Syndrome: A Modern Medical Mystery. *Newsweek, November 12,* 62-70.

Culver, Charles M. and Gert, Bernard (1982). *Philosophy in Medicine.* Oxford: Oxford University Press.

Curd, Patrician (Ed.) (1996). *A Presocratics Reader.* Indianapolis: Hackett Publishing Company.

Cutter, Mary Ann Gardell (1988). Explaining AIDS: A Case Study. In: E.T. Juengst and B. Koenig (Eds.), *The Meaning of AIDS: Perspectives From the Humanities* (pp. 21-29). New York: Praeger Scientific.

Cutter, Mary Ann Gardell (1992). Value Presuppositions in Diagnosis: A Case Study of Cervical Cancer. In:: J.L. Peset et al. (Eds.), *The Ethics of Diagnosis* (pp. 147-154). Holland: Kluwer Academic Publishers.

Cutter, Mary Ann Gardell (1993). Medicine: Explanation, Manipulation, and Creativity. In: C. Delkeskamp-Hayes and M.A.G. Cutter (Eds.), *Science, Technology, and the Art of Medicine* (pp. 251-158). Holland: Kluwer Academic Press.

Cutter, Mary Ann Gardell (1994). Technoscience Metaphors and Contemporary Genetics. *Voices, 1, 2,* 3-4.

Cutter, Mary Ann Gardell (1997). Engelhardt's Analysis of Disease: Implications for a Feminist Epistemology. In: B. Minogue et al. (Eds.), *Reading Engelhardt* (pp. 139-148). Holland: Kluwer Academic Publishers.

Cutter, Mary Ann Gardell (1998). Negotiating Diverse Values in a Pluralist Society: Limiting Access to Genetic Information. In: R.F. Weir (Ed.), *Ethical and Legal Implications of Stored Tissue Samples* (pp. 143-159). Iowa: University of Iowa Press, Iowa.

Cutter, Mary Ann Gardell (2002a). Ethics of Managed Care: An Excercise in te Challenges of Resolving Conflict. In: W. Bondeson and J. Jones (Eds.). *The Ethics of Managed Care* (pp. 127-138). Holland: Kluwer Academic Publishers.

Cutter, Mary Ann Gardell (2002b). Introduction. *The Journal of Medicine and Philosophy*, *27,3*,251-256.

Cutter, Mary Ann Gardell (2002c). Local Bioethics Discourse: In Search of Grounding. In: J. Po-wah (Ed.). *Cross-Cultural Perspectives on the (Im)possibility of Global Bioethics* (pp. 319-334). Holland: Kluwer Academic Publishers.

Cutter, Mary Ann Gardell (2002d). The Mind's Eye and the Hand's Brain: On Wartofsky's Philosophy of Medicine. In: C: Gould (Ed.), *Constructivism and Practice: Toward a Social and Historical Epistemology* (pp. 243-257). New York: Rowan and Littlefield.

Cutter, Mary Ann Gardell (2002e). Roman Catholic Moral Theology and the Allocation of Resources to Critical Care: Boundaries of Faith and Reason. In : H.T. Engelhardt, Jr. and M Cherry (Eds.), *Critical Care: A Roman Catholic Perspective on the Allocation of Scarse Resources* (pp. 310-319). Washington, D.C.: Georgetown University Press.

Cutter, Mary Ann Gardell and Shelp, Earl E. (Eds.) (1991). *When Are Competent Patients Incompetent?*, Holland: Kluwer Academic Publishers.

D'Amico, Robert (1995). Is Disease a Natural Kind?*The Journal of Medicine and Philosophy*, *20, 5*, 551-569.

Davis, Audrey B. (1971) Some Implications of the Circulation Theory and Treatment in the Seventeenth Century. *Journal of the History of Medicine, 26,* 28-39.

Davis, Audrey B. (1981). *Medicine and Its Technology: An Introduction to the History of Medical Instrumentation.* Connecticut: Greenwood Press.

Davis, Audrey B. and Appel, Toby (1979). *Bloodletting Instruments in the National Museum of History and Technology.* Washington, D.C.: Smithsonian Institution Press.

Delkeskamp-Hayes, Corinna and Cutter, Mary Ann Gardell (Eds.) (1993). *Science, Technology, and the Art of Medicine: European-American Dialogues.* Holland: Kluwer Academic Publisher.

Department of Health and Human Services (1985). *Women's Health: Report of the Public Service Task Force on Women's Issues*, Vol. 2. Washington, D.C.: Public Health Services.

DeRose, K. (1992) Contextualism and Knowledge Attribution. *Philosophy and Phenomenological Research, 52,* 913-929.

Descartes, René (1972 [1650]). *Treatise of Man*. Trans. T. S. Thomas. Massachusetts: Harvard University Press

Descartes, René (1993 [1641]). *Meditations of First Philosophy* 3rd ed. Trans. Donald A. Cress. Indiana: Hackett Publishing Company.

Dewey, John (1925). *Experience and Nature.* Illinois: Open Court.

Dilthey, Wilhelm (2002 [1910]). *Formation of the Historical World in the Human Sciences.* Ed. R.A. Makkreel and F. Rodi. New Jersey: Princeton University Press.

Dresser, Rebecca (1992). Wanted Single White Male for Medical Research. *The Hastings Center Report, January-February,* 24-29.

Dresser, Rebecca (1996). What Bioethics Can Learn from the Women's Health Movement. In: S.M. Wolf (Ed.), *Feminism and Bioethics: Beyond Reproduction* (pp. 144-159). New York: Oxford University Press.

Driesch, Hans (1914). *The History and Theory of Vitalism.* Trans. C.K. Ogden. London: Macmillan.

Duesberg, Peter H. (1987). Retrovirus as Carcinogens and Pathogens: Expectations and Reality. *Cancer Research, 47,* 1199-1210.

Duesberg, Peter H. (1988). HIV is Not the Cause of AIDS. *Science, 241, 4865,* 514, 517.

Duesberg, Peter H. (1989). Human Immunodeficiency Virus and Acquired Immunodeficiency Syndrome: Correlation But Not Causation. *Proceedings of the National Academy of Science, 86, 3,* 755-764.

Duesberg, Peter H. (1991). AIDS Epidemiology: Inconsistencies with Human Immunodeficiency Virus and With Infectious Disease. *Proceedings of the National Academy of Science, 88, 4,* 1575-1579

Duesberg, Peter H. (1994). Infectious AIDS--Stretching the Germ Theory Beyond Its Limits. *International Archives of Allergy and Immunology, 103, 2,* 118-127.

Dula, Annette and Goering, Sara (Eds.) (1994). *"It Just Ain't Fair": The Ethics of Health Care for African-Americans.* Connecticut: Praeger.

Duns Scotus, John (1983). *A Treatise On God As First Principle.* Illinois: Franciscan Herald.

Edelman, K. et al. (1991). HIV Does Not Cause AIDS--Impact of T.V. Programme on Attitudes to Zidovudine in HIV Patients. *VIIth Int Conf on AIDS* (abstract no. W.B. 2097), *June 16-21,* 45-53..

Ehrlich, Paul (1880). On the Immunity with Especial Reference to the Relations Existing Between the Distribution and the Action of Antigens. *Harben Lectures,* lecture 1, London.

Elton, G.R. (1991). *Return to Essentials: Some Reflections on the Present State of Historical Study* (esp. Ch. 3). England: Cambridge University Press.

Engel, George (1977). The Need for a New Medical Model: A Challenge for Biomedicine. *Science, 196,* 129-136.

Engelhardt, H. Tristram, Jr. (1974). Explanatory Models in Medicine: Facts, Theories, and Values. *Texas Reports on Biology and Medicine, 32, Spring,* 225-239.

Engelhardt, H. Tristram, Jr. (1977). Is There a Philosophy of Medicine? In: F. Suppe and P.D. Asquith (Eds.), *PSA 1976* (pp. 94-108). Michigan: Philosophy of Science Association.

Engelhardt, H.Tristram, Jr. (1981 [1974]). Disease of Masturbation. In: A.L. Caplan et al. (Eds.), *Concepts of Health and Disease: Interdisciplinary Perspectives,* (pp. 267-280). Massachusetts: Addison-Wesley Publishing Company.

Engelhardt, H. Tristram, Jr. (1981 [1975]). The Concepts of Health and Disease. In : A.L. Caplan et al. (eds.), *Concepts of Health and Disease* (pp. 31-46). Massachusetts: Addison-Wesley Publishing Company.

Engelhardt, H. Tristram, Jr. (1982). The Subordination of the Clinic. In: B. Gruzolski and C. Nelson (Eds.). *Value Conflicts in Health Care Delivery* (pp.41-57). Massachusetts: Ballinger.

Engelhardt, H. Tristram, Jr. (1984). Clinical Problems and the Concept of Disease. In: L. Nordenfelt and B.I.B. Lindahl (Eds.), *Health, Disease, and Clinical Explanation* (pp. 27-41). Holland: D. Reidel Publishing Company.

Engelhardt, H. Tristram, Jr. (1985). Typologies of Disease: Nosologies Revisited. In: K.F. Schaffner (Ed.), *Logic of Discovery and Diagnosis* (pp. 56-71), California: University of California Press.

Engelhardt, H. Tristram, Jr. (1986). Doctoring the Disease, Treating the Complaint, Helping the Patient:
    Some of the Works of Hygeia and Panacea. In: H.T. Engelhardt and D. Callahan (Eds.), *Knowing
    and Valuing: The Search for Common Roots* (pp. 225-249). New York: Hastings-on-Hudson.

Engelhardt, H. Tristram, Jr. (1996). *The Foundations of Bioethics*, 2nd ed. New York: Oxford University Press.

Engelhardt, H. Tristram, Jr., with Edmund L. Erde (1980). The Philosophy of Medicine. In: P.T. Durbin
    (Ed.), *A Guide to the Culture of Science, Technology, and Medicine* (pp. 336-465). New York:
    MacMillan.

Engelhardt, H.Tristram, Jr. and Cherry, Mark (2002). *Allocating Scarce Medical Resources: Roman Catholic
    Perspectives.* Washington, D.C.: Georgetown University Press.

Engelhardt, H. Tristram, Jr. and Spicker, Stuart F. (Eds.) (1975). *Evaluation and Explanation in the
    Biomedical Sciences.* Holland: D. Reidel Publishing Co.

Equal Employment Opportunity Commission (1995). *Compliance Manual*, Vol. 2. EEOC Order 915.002,
    Section 902, March 14.

Ermarth, Elizabeth Deeds (1992). *Sequel to History: Postmodernism and the Crisis of Representational Time.*
    New Jersey: Princeton University Press.

Ewing, Alfred C. (1947). *The Definition of Good.* New York: Macmillan Publishing Company.

Faber, Knud (1923). *Nosography in Modern Internal Medicine.* London: Oxford University Press.

Fabrega, Horacio, Jr. (1972). Concepts of Disease: Logical Features and Social Implications. *Perspectives
    in Biology and Medicine, 15,* 583-616.

Faden, Ruth et al. (1996). Women as Vessels and Vectors: Lessons from the HIV Epidemic. In: S. Wolf (Ed.),
    Feminism and Bioethics: Beyond Reproduction (pp. 266-278). New York: Oxford University Press.

Fauci, Anthony (1993). Multifactorial Nature of Human Immunodeficiency Virus Disease: Implications for
    Therapy. *Science, 3136,* 1011-1018.

Fausto-Sterling, Anne (1987). Society Writes Biology/Biology Constructs Gender. *Daedalus, 116, 4,* 61-76.

Fausto-Sterling, Anne (1992). *Myths of Gender: Biological Theories About Women and Men.* New York:
    Basic Books.

Fausto-Sterling, Anne (1996). The Five Sexes: Why Male and Female Are Not Enough. In: K.E. Rosenblum
    and T.C. Travis (Eds.). *The Meaning of Difference* (pp. 68-73). New York: McGraw-Hill
    Publishing Co.

Fausto-Sterling, Anne (2000). *Sexing the Body: Gender Politics and the Construction of Sexuality.* New
    York: Basic Books.

Feinstein, Alvan (1967). *Clinical Judgment.* Maryland: Williams and Wilkins.

Feldman, M. and MacCullough, M. (1971). *Homosexual Behaviour: Therapy and Assessment.* Oxford:
    Pergamon Press.

Fischer, Joannie (2000). Snipping Away at Human Disease. *U.S. News and World Reports, October 23,* 58-59.

Fleck, Ludwik (1979 [1935]). *Genesis and Development of a Scientific Fact*. Ed. T.J. Trenn and R.K. Merton. Trans. F. Bradley and T.J. Trenn. Chicago: University of Chicago Press.

Food and Drug Administration (1993). *Guidelines for the Study and Evaluation of Gender Differences in the Clinical Evaluation of Drugs*. Washington, D.C.: Food and Drug Administration.

Foot, Philippa (1959). Moral Beliefs. *Proceedings of the Aristotelian Society, 59*, 83-94.

Foucault, Michel (1972 [1969]). *The Archaeology of Knowledge*. Trans. A.M. Sheridan Smith. New York: Pantheon Books.

Foucault, Michel (1973 [1963]). *The Birth of the Clinic: An Archeology of Medical Perception*. Trans. by A.M. Sheridan Smith. New York: Pantheon Books.

Fox, Lynda (1995). New Technologies, New Dilemmas: S.B. 94-058 Limits Use of Genetic Testing Information. *The Colorado Lawyer, 24, 2,* 275-276.

Fox, Michael W. (2001). *Bringing Life to Ethics: Global Bioethics for a Humane Society*. New York: State University of New York Press.

Fox-Keller, E. (1991). Genetics, Reductionism, and Normative Uses of Biological Information. *Southern California Law Review, 65*, 285-291.

Fracatoro, Girolamo (1530). *Syphilis sive Morbus Gallicus*. Venice.

Freeman, William L. (1998). The Role of Community in Research with Stored Tissue Samples. In: R.F. Weir (Ed.), *Stored Tissue Samples: Ethical, Legal, and Public Policy Implications* (pp. 267-301). Iowa: University of Iowa Press.

Frerichs, R.R. (1994). Personal Screening for HIV in Developing Countries. *Lancet, 343*, 960-962.

Freud, Sigmund (1962 [1905]). *Three Essays on the Theory of Sexuality*. Trans., and Ed. J. Strachey. London: Hogarth Press.

Freud, Sigmund (1966). *The Complete Introductory Lectures on Psychoanalysis*. Trans, and Ed. J. Strachey. New York: W. W. Norton.

Fuller, Steve (1995). Interdisciplinary Rhetoric: Lessons for Both Rhetor and Rhetorician. *Social Epistemology, 9 , 2,* April-June, 201-204.

Gadamer, H.G. (1992). *Truth and Method*. Trans. J. Weinsheimer and D. Marshall. New York: Crossroads.

Galilei, Galileo (1953 [1632]). *Dialogue Concerning The Two Chief World*. Trans. S. Drake. California: University of California Press.

Gallo, Robert C., and Montagnier, Luc (1987). The Chronology of AIDS Research. *Nature, 326, 6112,* 435-436.

Gallo, Robert C. et al. (1984). Frequent Detection and Isolation of Cytopathic Retroviruses (HTLV-III) From Patients with AIDS and At Risk for AIDS. *Science, 224,* 500-503.

Gardell, Stephen et al. (1990). Vampire Bat Saliva Plasminogen Activator Is Quiescent in Human Plasma in the Absence of Fibrin Unlike Human Tissue Plasminogen Activator. *Blood, 76, 12,* 2560-2564.

Gardell, Stephen et al. (1991), Effective Thrombolysis Without Marked Plasminema After Bolus Intravenous
      Administration of Vampire Bat Salivary Plasminogen Activator in Rabbits. *Circulation, 84 , 1,
      July,* 244-253.

Gardner, Howard (1993) *Multiple Intelligences: The Theory in Practice.* New York: Basic Books.

Garrison, Fielding H. (1929). *An Introduction to the History of Medicine* 4th ed.. Pennsylvania: W.B.
      Saunders Company.

Gavaret, Jules (1840). *Principes Généraux de Statistique Médicale.* Paris. Trans. By H. Wulff et al. (1986),
      Philosophy of Medicine: An Introduction (pp. 35-36). Oxford: Blackwell Scientific Publications.

Goldsmith, M.F. (1992). Specific HIV-Related Problems of Women Gain More Attention at a Price Affecting
      More Women. *Journal of the American Medical Association, 268,* 1814-1816.

Gonella, J.S. et al. (1984). Staging of Disease: A Case-Mix Measurement. *Journal of the American Medical
      Association, 251,* 637-644.

Goodwin, J.S. and Goodwin, J.M. (1984). The Tomato Effect: Rejection of Highly Efficacious Therapy.
      Journal of the American Medical Association, 251, 2387-2390.

Goosens, William K. (1980). Values, Health, and Medicine. *Philosophy of Science, 47,* 102.

Greenspan, Miriam (1993). *A New Approach to Women and Therapy.* Pennsylvania: TAB Books.

Grene, Marjorie (1974). *The Knower and the Known.* California: University of California.

Grene, Marjorie (1977). Philosophy of Medicine: Prolegomena to a Philosophy of Science. In P. Asquith and
      F. Suppe (Eds.), *PSA 1976* (pp. 77-93). Michigan: Philosophy of Science Association.

Grene, Marjorie (1978). Individuals and Their Kinds: Aristotelian Foundations of Biology. In : S.F. Spicker
      (Ed.), *Organism, Medicine, and Metaphysics* (pp. 121-136). Holland: D. Reidel Publishing
      Company.

Grmek, M.D. (1990). *History of AIDS. Emergence and Origin of a Modern Pandemic.* Trans. R. Maulitz
      and J. Duffin. New Jersey: Princeton University Press.

Gutting, Gary (1980). *Paradigms and Revolutions: Appraisals of Thomas Kuhn's Philosophy of Science.*
      Indiana: University of Notre Dame Press.

Habermas, Jürgen (1972). *Knowledge and Human Interests.* Trans. Jeremy J. Shapiro. Boston: Beacon Press.

Hanson, N.R. (1958). *Patterns of Discovery.* Cambridge: Cambridge University Press.

Hardy, A.M. et al. (1986). The Economic Impact of the First 10,000 Cases of Acquired Immunodeficiency
      Syndrome in the United States. *Journal of the American Medical Association, 255,* 209-215.

Haré, Richard M. (1952) *The Language of Morals.* England: Oxford University Press.

Harlan, David (1989). Intellectual History and the Return of Literature. *American Historical Review, 94,* 581-
      609.

Harrison, G.A. et al (2002). *Human Biology: An Introduction to Human Evolution, Variation, Growth, and
      Adaptability.* 3rd ed. New York: Oxford University Press.

Hartl, D.L. et al. (2002). *Essential Genetics: A Genomics Perspective*. 3rd ed. Boston: Jones and Bartlett.

Health Insurance Portability and Accountability Act (1996). Public Law 104-191.

Hegel, G.W.F. (1955). *Hegel's Lectures on the History of Philosophy*. Trans. E.S. Haldane and F.H. Simson, 3 vols. New York.

Hegel, G.W.F.: 1970 (1830), *Philosophy of Nature*. Trans. M.J. Petry. London: Allen and Unwin.

Heidegger, Martin (1966 [1927]). *Being and Time*. Trans. J. Stambaugh. New York: State University of New York Press.

Hempel, Carl (1965). *Aspects of Scientific Explanation*. New York: Free Press.

Hempel, Carl G. and Oppenheim, Paul (1970 [1948]). Studies in the Logic of Explanation. In: B.A. Brody (Ed.). *Readings in the Philosophy of Science*, pp. 8-27. New Jersey: Prentice Hall.

Hill, S. (1995). Neurological and Clinical Markers for a Severe Form of Alcoholism in Women. *Alcohol Health and Research World, 19, 13,* 249-256.

Hippocrates (1923), *Hippocrates and the Fragments of Heraclitus*, 4 Vols., Trans. W.H.S. Jones. Massachusetts: Harvard University Press.

Hippocrates (1943). *Regimen*. Trans. W H.S. Jones. Cambridge: Harvard University Press.

Hobbes, Thomas (1994 [1651]) *Leviathan*. Ed. E. Curley. Indiana: Hackett Publishing Company.

Hoffmann, Friedrich (1971 [1695]). *Fundamente Medicinae*. Trans. L. King. London: MacDonald.

Holtzman, N.A. (1970). Dietary Treatment of Inborn Errors of Metabolism. *Annual Review of Medicine, 21,* 115-132.

Holtzman, M.A. (1989). *Proceed With Caution: Predicting Genetic Risks in the Recombinant DNA Era*. Maryland: Johns Hopkins University Press.

Horkheimer, Max (1982). *Critical Theory: Selected Essays*. Trans. M.J. O'Connell. New York: Continuum.

Hubbard, Ruth (1995). Rethinking Women's Biology. In: P.S. Rothenberg (Ed.). *Race, Class, Gender in the United States,* pp. 32-33. New York: St. Martin's.

Hudson, Kathy et al. (1995). Genetic Discrimination and Health Insurance: An Urgent Need for Reform. *Science, 27,* 391-393.

Hudson, Robert P. (1983). *Disease and Its Control: The Shaping of Modern Thought*. Connecticut: Greenwood Press.

Hume, David (1938 [1740]). *An Abstract of a Treatise of Human Nature*. Ed. J.M. Keynes and P. Sraffa. Cambridge: Cambridge University Press.

Hume, David (1980 [1739-1740]). *Treatise of Himan Nature*. New York: Oxford University Press.

Husserl, Edmund (1970). *The Crisis of European Sciences and Transcendental Phenomenology: An Introduction to Phenomenological Philosophy*. Trans. D. Carr. Illinois: Northwestern University Press.

Illich, Ivan (1976). *Medical Nemesis: The Expropriation of Health*. New York: Pantheon Books.

Jennings, D. (1986). The Confusion Between Disease and Illness in Clinical Medicine. *Canadian Medical Association Journal, 135*, 865-880.

Johannsen, Wilhelm (1909). *Elemente der exakten Erblichkeitslehre*. Jena: Gustav Fischer.

Jonas, Wayne B. (1993). Evaluating Unconventional Medical Practices. *Journal of NIH Research, 5*, 64-66.

Jones, James (1981). *Bad Blood*. New York: The Free Press.

Jones, W.H.S. (1946). *Philosophy and Medicine in Ancient Greece*. Maryland: Johns Hopkins University Press.

Jonsen, Albert R. (1998). *The Birth of Bioethics*. New York: Oxford University Press.

Juengst, Eric T. (1995a). The Ethics of Prediction: Genetic Risk and the Physician-Patient Relationship. *Genome Science and Technology, 1, 1*, 21-36.

Juengst, Eric T. (1995b). Prevention and the Goals of Genetic Medicine. *Human Gene Therapy, 6* , December, 1595-1605.

Kant, Immanuel (1929 [1789]). *Critique of Pure Reason*. Trans. N.K. Smith. New York: St. Martin's Press.

Kant, Immanuel (1985 [1785]). *Foundations of the Metaphysics of Morals*. New York: Macmillan Publishing Company.

Kant, Immanuel (1987 [1790]). *Critique of Judgment*. Trans. And Intro. W.S. Pluhar. Foreward by M. Gregor. Indiana: Hackett Publishing Company.

Kaufman, James M. (2001). *Characteristics of Emotional and Behavioral Disorders of Childhood and Youth*. Ohio: Merrill Prentice Hall.

Kass, Leon (1981 [1975]). Regarding the End of Medicine and the Pursuit of Health. In: A.L. Caplan et al. (Eds.), *Concepts of Health and Disease* (pp. 3-30). Massachusetts: Addison-Wesley Publishing Company.

Kendell, R. (1975). *The Role of Diagnosis in Psychiatry*. Oxford: Blackwell Scientific Publishers.

Kendell, R. (1979). Alcoholism: A Medical or a Political Problem. *British Medical Journal*, 1, 367-371.

Khushf, George (1995). Expanding the Horizon of Reflection on Health and Disease. *Journal of Medicine and Philosophy, 20, 5*, 461-473.

Khushf, George (1997). Why Bioethics Needs the Philosophy of Medicine: Some Implications of Reflection on Concepts of Health and Disease. *Theoretical Medicine and Bioethics, 18*, 145-163.

Khushf, George (1999). The Aesthetics of Clinical Judgment: Exploring the Link Between Diagnostic Elegance and Effective Resource Utilization. *Medicine, Health Care, and Philosophy, 2*, 141-159.

King, Lester S. (1966). Boissier de Sauvages and 18th Century Nosology. *Bulletin of the History of Medicine, 15*, 43-51.

King, Lester S. (1970) Empiricism and Rationalism in the Works of Thomas Sydenham. *Bulletin of the History of Medicine, 19*, 1-11.

King, Lester S. (1975). Some Basic Explanations of Disease: An Historical Viewpoint. In: H.T. Engelhardt, Jr. and S.F. Spicker (Eds.). *Evaluation and Explanation in the Biomedical Sciences* (pp. 11-27). Holland: D. Reidel Publishing Company.

King, Lester S. (1981 [1954]). What is Disease? In: A.L. Caplan et al. (Eds.), *Concepts of Health and Disease: Interdisciplinary Perspectives* (pp. 193-203). Massachusetts: Addison-Wesley Publishing Company.

King, Lester S. (1982). *Medical Thinking: A Historical Preface*. New Jersey: Princeton University Press.

Kinsey, A. (1953). *Sexual Behavior in the Human Female*. Philadelphia: W.B. Saunders.

Klatzmann, D. et al. (1984). Selective Tropism of Lymphadenopathy Associated Virus (LAV) for Helper Inducer T. Lymphocytes. *Science, 225, 4657*, 59-63.

Kleiner, Carolyn (2001). The GEDs New Math. *U.S. News and World Report, December 17*, 42-43.

Koch, Robert (1932 [1882]). *The Aetiology of Tuberculosis*. New York: National Tuberculosis Association.

Kraupl-Taylor, F. (1979). *The Concepts of Illness, Disease, and Morbus*. Cambridge: Cambridge University Press.

Kuhn, Thomas S. (1970 [1962]. *The Structure of Scientific Revolutions*. 2nd ed.. Illinois: University of Chicago Press.

La Mettrie, Julien-Ofray (1961 [1748]). *Man as a Machine*. Illinois: Open Court.

Laennec, Theophile-Hyacinthe (1834 [1819]). *Treatise on Medical Ausculation and the Diseases of the Lungs and Heart*. London, 1834.

Lafaille,Robert and Fulder, Stephen (1993). *Towards a New Science of Health*. London: Routledge.

Lain-Entralgo, Pedro (1970). *Therapies of the Word in Classical Antiquity*. Connecticut: Yale University Press.

Lakatos, Imre (1970). Falsification and the Methodology of Scientific Research Programmes. In : I. Lakatos and A. Musgrave (Eds.), *Criticisms and the Growth of Knowledge* (pp. 91-125). England: Cambridge University Press.

Landers, Eric S. and Schork, Nicholas J. (1994). Genetic Dissection of Complex Traits. *Science, 265*, 2037-2048.

Laor, Nathanial and Agassi, Joseph (1990). *Diagnosis: Philosophical and Medical Perspectives*. Holland: Kluwer Academic Publishers.

Lasagna, Louis (1962). *The Doctor's Dilemma* Gollanez, London, pp. 3-21.

Laudan, Larry (1977). *Progress and Its Problems: Toward a Theory of Scientific Growth*. California: University of California Press.

Laufer, M. (1957). Hyperkinetic Impulse Disorders in Children's Behaviour Problems. *Psychological Medicine, 18*, 38-49.

Lennox, J.G. (1995). Health as an "Objective Value". *Journal of Medicine and Philosophy, 20*, 499-511.

Leshner, Alan I. (1997). Addiction Is A Brain Disease, and It Matters. *Science, 278*, 45-7.

Levy, J.A. et al. (1984). Isolation of Lymphocytopathic Retroviruses from San Francisco Patients with AIDS. *Science, 225*, 840-842.

Lewis , C.I. (1946). *An Analysis of Knowledge and Valuation.* Illinois:

Lewontin, Richard (1991). *Biology as Ideology: The Doctrine of DNA,.* New York: Harper Perennial.

Li, F.P. et al. (1992). Recommendations on Predictive Testing for Germ-Line p53 Mutations Among Cancer-Prone Individuals. *Journal of the National Cancer Institute, 84*, 1156-1160

Lindahl, Ingemar and Nordenfelt, Lennart (Eds.) (1984). *Health, Disease, and Causal Explanations in Medicine.* Holland: D. Reidel Publishing Company. •

Lindeboom, G.A. (1968). *Hermann Boerhaave.* London: Mentheun and Co., Ltd.

Lindeboom, G.A. (1979). *Descartes and Medicine.* Amsterdam: Rodopi N.V.

Linnaeus, Carolus (1763). *Genera morborum, in auditorum usum.* Upsaliae: Steinert.

Lobkowicz, Nicholas (1967). *Theory and Practice: History of a Concept from Aristotle to Marx.* Indiana: University of Notre Dame.

Locke, John (1975 [1690]). *An Essay Concerning Human Understanding.* Ed. by P.H. Nidditch. New York: Oxford University Press.

Louis, Pierre-Charles-Alexandre (1835). *Recherches sur lew Effets de la Saignée dans Quelques maladies Inflammatoires.* Paris.

Lund, Fred B. (1936). *Greek Medicine.* New York: Hoeber.

Lurie, Nicole et al. (1993). Preventive Care for Women: Does the Sex of the Physician Matter? *New England Journal of Medicine, 329, 7,* 478-482.

Lyotard, Francois (1984 [1979]), *The Postmodern Condition: A Report on Knowledge.* Trans. G. Bennington and B. Massumi. Minneapolis: University of Minnesota Press.

Macgregor, Frances M. (1960). *Social Science in Nursing: Applications for the Improvement of Patient Care.* New York: Russell Sage Foundation.

MacIntyre, Alasdaire (1981). *After Virtue.* Indiana: Notre Dame University Press.

Mackie, J.L. (1965). Causes and Conditions. *American Philosophical Quarterly 2*, 245-264.

Mahowald, Mary (1999). *Genes, Women, and Equality.* New York: Oxford.

Mahowald, Mary (2000). *Behavioral Genetics and Alcohol Consumption in Women.* New York: Oxford University Press.

Mappes, Thomas and Zembaty, Jane F. (1997). *Social Ethics.* New York: McGraw Hill.

Margolis, Joseph (1976). The Concepts of Disease. *The Journal of Medicine and Philosophy, 1,* 238-255.

Marx, Karl (1961 [1844]). *Economic and Philosophic Manuscripts of 1844.* Trans. M. Milligan. Moscow: Foreign Languages Publishing House.

Maudsley, H. (1868). Illustrations of a Variety of Sanity. *Journal of Medical Science, 14*, 149-156.

Mayr, Ernst (1982) *The Growth of Biological Thought*. Massachusets: Harvard University Press.

Mayr, Ernst (1988). *Toward a New Philosophy of Biology*. Massachusetts: Belknap Press.

McCray, E. et al. (1986). Occupational Risk of the Acquired Immunodeficiency Syndrome Among Health Care Workers. *New England Journal of Medicine, 314,* 1127-1132.

McCullough, Laurence B. (1998). *John Gregory and the Invention of Professional Medical Ethics and Profession of Medicine*. Holland: Kluwer Academic Publishers.

McElhinney, Thomas K. (Ed.) (1981). *Human Values Teaching Program for Health Professionals*. Pennsylvania: Whitmore Publishing Company.

McInerney, Joseph D. (2002). Issues in Genetic Education. *Journal of Medicine and Philosophy, 27 , 3*, 369-390.

McInerney, Joseph D. and Moore, Randy (1993). Voting in Science: Raise Your Hand If You Want Humans to Have 48 Chromosomes. *The American Biology Teacher, 55, 3*, 132-133.

McKay, Doug (1983). *Asylum of the Gilded Pill: The Story of Cragmor Sanatorium*. Colorado: State Historical Society of Colorado.

McMullin, Ernan (1979). A Clinician's Quest for Certainty. In: H.T. Engelhardt, Jr. et al. (eds.), *Clinical Judgment* (pp. 115-130). Holland: D. Reidel Publishing Company.

McMullin, Ernan (1987). Scientific Controversies and Its Termination. In: H.T. Engelhardt, Jr. and A.L. Caplan (eds.), *Scientific Controversies: Case Studies in the Resolution and Closure of Disputes in Science and Technology* (pp. 49-92). England: Cambridge University Press.

Meinert, Curtis (1995). The Inclusion of Women in Clinical Trials. *Science , 269*, 795-796.

Meinong, Alexius (1983). *On Assumptions*. Ed. And trans. J. Heanue. California: University of California Press.

Merton, Vanessa (1996). Ethical Obstacles to the Participation of Women in Biomedical Research. In: S. Wolf (Ed.), *Feminism and Bioethics: Beyond Reproduction* (pp. 216-251). New York: Oxford University Press.

Mill, John Stuart (1874). *A System of Logic*. 8th ed. New York: Harper and Brothers.

Mill, John Stuart (2002 [1861]). *Utilitarianism*. 2nd ed. Ed. G. Sher. Indiana: Hackett Publishing Company.

Montagnier, Luc et al. (1984). A New Lymphotropic Retrovirus: Characterization and Possible Role in Lymphadenopathy and Acquired Immune Deficiency Syndrome. In : R.C. Gallo et al. (eds.), *Human T-Cell Leukemia/Lymphoma Virus* (pp. 363-379). New York: Cold Springs Harbor Laboratory.

Moore, G.E. (1903). *Principia Ethica*. England: Cambridge Unievrsity Press.

Moore, J.A. (1992). *Science as a Way of Knowing: The Foundations of Modern Biology*. Massachusetts: Harvard University Press.

Mordacci, Roberto (1995). Health as an Analogical Concept. *Journal of Medicine and Philosophy*, *20*, *5*, 475-497.

Morgagni, G.B. (1981 [1761], 'The Author's Preface', *The Seats and Causes of Disease*, Wells and Lilly, Boston. In: A.L. Caplan et al (Eds.), *Concepts of Health and Disease* (pp. 157-165). Massachusetts: Addison-Wesley Publishing Company.

Morton, Richard (1720). *Phthisiologia: Or A Treatise of Consumptions.* 2nd ed. London: W. and J. Innys.

Moss, Lenny (2003). *What Genes Can't Do.* Massachusetts: MIT Press.

Munson, Ronald (2000). *Intervention and Reflection: Basic Issues in Medical Ethics.* 6th ed., California: Wadsworth.

Murphy, Edmond (1976). *The Logic of Medicine.* Maryland: Johns Hopkins University Press.

Murphy, Timothy F. (1994). *Ethics in an Epidemic: AIDS, Morality, and Culture.* California: University of California Press.

Nash, Mark (2000). The Simple Solution to the AIDS Epidemic. *The Philadelphia Trumpet, July*, 11-13.

National Academy of Science (1998). *Teaching About Evolution and The Nature of Science.* Washington, D.C.: National Academy Press.

National Coalition for Health Professional Education in Genetics (2002). *Genetics and Major Psychiatric Disorders: A Program for Genetic Counselors.* Maryland: National Coalition for Health Professional Education in Genetics.

National Commission for the Study of Ethical Problems in Medicine and Biomedical and Behavioral Research (1983). *Making Health Care Decisions.* Washington, D.C.: U.S. Government Printing Office.

National Institute of Allergy and Infectious Diseases, National Institutes of Health (1995). *The Relationship Between the Human Immunodeficiency Virus and the Acquired Immunodeficiency Syndrome.* Maryland: National Institutes of Health.

National Institutes of Health (1993). *NIH Revitalization Act.* Washington, D.C.: National Institutes of Health.

National Institutes of Health/Department of Energy (1992). *Task Force on Genetic Information and Insurance.* Washington, D.C.: U.S. Government Printing Office.

National Institutes of Health et al. (1999a). *Cell Biology and Cancer.* Washington, D.C.: NIH Publication (99-4646).

National Institutes of Health et al. (1999b). *Emerging and Re-emerging Infectious Diseases.* Washington, D.C.: NIH Publication (99-4645).

National Institutes of Health et al.(1999c). *Human Genetic Variation.* Washington, D.C.: NIH Publication (99-4647).

National Public Radio (2002). July 24.

Neese, R.M. and Williams, George (1994). *Why We Get Sick: The New Science of Darwinian Medicine.* New York: Times Books.

Neese, R.M., and Williams, George (1998). Evolution and the Origins of Disease. *Scientific America, 280 ,* 86-93.

Nelkin, D. and Lindee, M.S. (1995). *The DNA Mystique: The Gene as a Cultural Icon.* New York: W.H. Freeman.

Newton, Isaac (1999 [1687]). *The Principia: Mathematical Principles of Natural Philosophy.* Trans. I.B. Cohen and A. Whitman. California: University of California Press.

Niebyl, Peter H. (1971). Sennert, van Helmont, and Medical Ontology. *Bulletin of the History of Medicine, 45,* 115-137.

Nordenfelt, Lennart (1987). *On The Nature of Health: An Action-Theoretic Approach.* Holland: D. Reidel Publishing Company. See especially Appendix, pp. 151-173.

Nordenfelt, Lennart (2001). *Health, Science, and Ordinary Language.* New York: Rodolpi.

Nordenfelt, Lennart and B. Ingemar B. Lindahl (Eds.) (1984). *Health, Disease, and Causal Explanations in Medicine.* Holland: D. Reidel Publishing Company.

Northrup, Christine (1994). *Women's Bodies, Women's Wisdom: Creating Physical and Emotional Health and Healing.* New York: Bantam Books.

Novick, Peter (1988). *The Noble Dream: The "Objectivity Question" and the American Historical Profession.* England: Cambridge University Press.

Ogden, C.K. and Richards, I.A. (1956). *The Meaning of Meaning* 8th Ed. New York: Harcourt Brace and Co.

Organization for Economic Cooperation and Development (2000). *Health Data 2000.* Paris: OECD.

Osborne, Lawrence (2001). Royal Disturbances. *The New York Times Magazine,* 98-102.

Overall, Christine (1995). *Perspectives on AIDS: Ethical and Social Issues.* New York: Oxford University Press.

Owens, Douglas et al. (1996). Cost-Effectiveness of HIV Screening in Acute Care Settings. *Archives of Internal Medicine, 156,* 394-404.

Owens, Joseph (1977). Aristotelian Ethics, Medicine, and the Changing Nature of Man. In: S.F. Spicker and H.T. Engelhardt, Jr. (Eds.), *Philosophical Medical Ethics: Its Nature and Significance* (pp. 127-142). Holland: D. Reidel Publishing Company.

Pagel, Walter (1972). Van Helmont's Conception of Disease--To Be or Not To Be? The Influence of Paracelsus. *Bulletin of the History of Medicine, 46,* 419-453.

Pantaleo, G. et al. (1993). The Immunogenesis of Human Immunodeficiency Virus Infection. *New England Journal of Medicine, 328, 5,* 327-335.

Parsons, Talbott (1958). Definitions of Health and Illness in the Light of American Values and Social Structures. In E.G. Jaco (Ed.), *Patients, Physicians, and Illness* (pp. 167-187). Illinois: The Free Press.

Pascal, Blaise (1995 [1670]). *Pensées.* Trans, and with an introduction by A.J. Kailsheimer. New York: Penguin.

Pasteur, Louis (1942 [1862]). *Memoirs on the Organic Corpuscles Which Exist in the Atmosphere*. In: L.
    Clandenning (ed.) and D.H. Clandenning (trans.), *Source Book of Medical History*. New York:
    Dover.

Peck, M. Scott (1978). *The Road Less Traveled*. New York: Simon and Schuster.

Pellegrino, Edmund D. (1976). Philosophy of Medicine: Problematic and Potential. *The Journal of Medicine
    and Philosophy, 1* , March, 5-31.

Pellegrino, Edmund D. (1979). The Anatomy of Clinical Judgments: Some Notes on Right Reason and Right
    Action. In: H.T. Engelhardt, Jr. et al. (Eds.), *Clinical Judgment: A Critical Appraisal* (pp. 169-
    194). Holland: D. Reidel Publishing Company.

Pellegrino, Edmund D. (1983). The Healing Relationship: The Architectonic of Clinical Medicine. In E.E.
    Shelp (ed.), *The Clinical Encounter: The Moral Fabric of the Patient-Physician Relationship* (pp.
    XXX-XXX). Holland: D. Reidel Publishing Company.

Pellegrino, Edmund D. (1998). What the Philosophy *of* Medicine *Is. Theoretical Medicine and Bioethics*,
    19, 315-336.

Pellegrino, Edmund D., and Thomasma, David C. (1981). *A Philosophical Basis of Medical Practice*. New
    York: Oxford University Press.

Pellegrino, Edmund D., and Thomasma, David C. (1988). *For the Patient's Good: A Restoration of
    Beneficence in Health Care*. New York: Oxford University Press.

Pence, Gregory (2000). *Classic Cases in Medical Ethics: Accounts of Cases That Have Shaped Medical
    Ethics*, 3rd ed. Boston: McGraw-Hill.

Peppin, John (1999). Business Ethics and Health Care: The Reemerging Institution-Patient Relationship.
    The Journal of Medicine and Philosophy, 24, 5, 535-550.

Perry, Ralph Barton (1954). *Realms of Value: A Critique of Human Civilization*. Massachusetts: Harvard
    University Press.

Pinel, Phillipee (1798). *Nosographie philosophique, ou la mèthode l'analse appliquè à la mèdicine*. Paris:
    Richard Caile et Ravier.

Plato (1992). *The Republic*. 2nd ed. Trans. G.M.A. Grube. Rev. C.D.C. Reeve. Indiana: Hackett Publishing
    Company.

Po-wah, Julia Tao Lai (Ed.) (2002). *Cross-Cultural Perspectives on the (Im)Possibility of Global Bioethics*.
    Holland: Kluwer Academic Publishers.

Poincaré, Henri (1996 [1908]). *Science and Method*. Bristol: Thoemmes Press.

Pope, Victoria (1996). Mad Russians. *U.S. News and World Report, December 16*, 38-43.

Popovic, M. et al. (1984). Detection, Isolation, and Continuous Production of Cytopathic Retroviruses
    (HTLV-III) from Patients with AIDS and Pre-AIDS. *Science, 224, 4648*, 497-500.

Popper, Karl (1959). *The Logic of Scientific Discovery*. London: The Free Press.

Popper, Karl (1965). *Conjectures and Refutations* 2nd ed. London: Routledge and Kegan Paul.

Portin, Petter (1993). The Concept of the Gene: Short History and Present Status. *Quarterly Review of Biology* 68, 2, 173-223.

Portin, Petter (2002). Historical Development of the Concept of the Gene. *The Journal of Medicine and Philosophy, 27, 3,* 257-286.

President's Commission for the Study of Ethical Problems in Medicine and Biomedical and Behavioral Research (1981). *Protecting Human Subjects.* Washington, D.C.: U.S. Government Printing Office.

Preston, N. (1983). Is Stuttering a Disease? *Journal of Pediatrics, 41,* 135-156.

Proctor, Robert (1988). *Racial Hygiene: Medicine Under the Nazis.* Cambridge: Harvard University Press.

Profet, Margie (1993). Menstruation as a Defense Against Pathogens Transported by Sperm. *The Quarterly Review of Biology, 68, 3,* 335-381.

Purdy, Laura (1996). in S. Wolf (Ed.), *Feminism and Bioethics: Beyond Reproduction* (pp. 65-79). New York: Oxford University Press.

Rather, L.J. (1959). Towards a Philosophical Study of the Idea of Disease. In C.M. Brooks and P.F. Cranefield (Eds.). *The Historical Development of Physiological Thought* (pp. 360-373). New York: Hafner.

Rather, L. et al. (1985). HTLV-III, LAV, ARV Are All Variants of Same AIDS Virus. *Nature, 313, 6004,* 636.

Reich, Warren T.(Ed.) (1995). *Encyclopedia of Bioethics,* rev. ed. New York: Macmillan Library Reference.

Reiser, Stanley J. (1978). *Medicine and the Reign of Technology.* England: Cambridge University Press.

Reznek, Lawrie (1987). *The Nature of Disease.* London: Routledge and Kegan Paul.

Reznek, Lawrie (1995). Dis-ease about Kinds: Reply to D'Amico. *The Journal of Medicine and Philosophy 20, 5,* 571-584.

Richart, Ralph M. (1976). Cervical Intraepithelial Neoplasia and The Cervicologist. *Canadian Journal of Medical Technologists, 38,* 1877-1880.

Roberts, Dorothy (1996). Women and Medicine. In: S. Wolf (Ed.), *Feminism and Bioethics: Beyond Reproduction.* New York: Oxford University Press.

Robinson, Daniel F. (1976). *The Enlightened Machine: An Analytical Introduction to Neuropsychology.* Calfornia: Dickenson Publishing Company.

Robinson, Daniel F. (1995). *An Intellectual History of Psychology.* 3rd ed. Wisconsin: University of Wisconsin Press.

Romanell, Patrick (1974). *John Locke and Medicine.* New York: Prometheus.

Rosenberg, Charles E. (1979). The Therapeutic Revolution: Medicine, Meaning, and Social Change in 19th Century America. In: J. Vogel and C.E. Rosenberg (eds.). *The Therapeutic Revolution: Essays in the Social History of American Medicine* (pp. 3-25). Pennsylvania: University of Pennsylvania Press.

Ross, W.D. (1930). *The Right and the Good.* New York: Oxford University Press.

Rothenberg, Paula S. (2001). *Race, Class, and Gender in the United States: An Integrated Study*, 5th ed. New York: Worth Publishers.

Roussea, F. et al. (1990). Direct Diagnosis by DNA Analysis of the Fragile X Syndrome of Mental Retardation. *New England Journal of Medicine, 325*, 1673-1681.

Sanchez-Gonzalez, Miguel A. (1990). Medicine in John Locke's Philosophy. *Journal of Medicine and Philosophy 15*, 675-695.

Sande, M. (1986). The Case Against Causal Transmission. *New England Journal of Medicine, 314*, 380-382.

Sartre, Jean-Paul (1969). *Nausea*. Trans. L. Alexander. Connecticut: New Directions Paperback.

Sassower, Raphael (1993). *Knowledge Without Expertise: On The Status of Scientists*. New York: State University of New York.

Sassower, Raphael (1995). *Cultural Collisions: Postmodern Technoscience*. New York: Routledge.

Sauvages de la Croix, Francois Boissier de (1768). *Nosologia methodica sistens morborum classes juxta Sydenhami mentem et botanicorum ordinem*, 5 Vols. Amsterdam: Fratrum de Tournes.

Schaffner, Kenneth F. (1980). Theory Structure in the Biomedical Sciences. *Journal of Medicine and Philosophy, 5*, 59-97.

Schaffner, Kenneth F. (1981). Causation and Responsibility: Medicine, Science, and Law. In: S.F. Spicker et al., *The Law-Medicine Relation: A Philosophical Exploration* (pp. 95-122). Holland: D. Reidel Publishing Company.

Schaffner, Kenneth (1993). *Discovery and Explanation in Biology and Medicine*. Illinois: University of Chicago.

Schaffner, Kenneth F. and Engelhardt, Jr., H. Tristram (1998). Medicine, Philosophy of. *Routledge Encyclopedia of Philosophy* Vol. 6 (pp. 264-269). London: Routledge.

Schaudin, Fritz (1905). *Arb. a.d.k. Gesundheit samte*, Berlin, xxii, 527-534.

Schmitt, Barton D. (1987). *Your Child's Health*. New York: Bantam Books.

Scriven, Michael (1959). Explanation and Prediction in Evolutionary Theory. *Science, 130*, 477-483.

Scriver C.R. and Waters, P.J. (1999). Monogenic Traits are Not Simple. *Trends in Genetics, 15, 7*, 267-272.

Sedgwick, Peter (1981 [1973]). Illness–Mental and Otherwise. In: A.L. Caplan et al. (Eds.). *Concepts of Health and Disease: Interdisciplinary Perspectives* (pp. 119-130). Massachusetts: Addison-Wesley Publishing Company.

Sedgwick, Peter (1982). *Psychopolitics*. New York: Harper and Row.

Seldin, Donald (1977). The Medical Model: Biomedical Sciences as the Basis for Medicine. *Beyond Tomorrow*. New York: Rockefellow University Press.

Selik, R.M. et al. (1984). Acquired Immune Deficiency Syndrome (AIDS) Trends in the United States, 1978-1982. *American Journal of Medicine, 76*, 493-500.

Shaffer, Jerome (1975). Roundtable Discussion. In H.T. Engelhardt, Jr., and S.F. Spicker (eds.), *Evaluation and Explanation in the Biomedical Sciences* (pp. 215-219). Holland: D. Reidel Publishing Company.

Shea, John (2000). *Spirituality and Health Care.* Illinois: Park Ridge Center.

Sherwin, Susan (1992). *No Longer Patient: Feminist Ethics and Health Care.* Pennsylvania: Temple University Press.

Sherwin, Susan (2001). Feminist Ethics and the Metaphor of AIDS. *Journal of Medicine and Philosophy*, *26 4*, 343-364.

Shilts, Randy (1987). *And the Band Played On: Politics, People, and the AIDS Epidemic.* New York: St. Martin's Press.

Shuval, Judith T. (1981). Contribution of Psychological and Social Phenomena to an Understanding of the Aetiology of Disease and Illness. *Social Science and Medicine, 15A*, 337-342.

Siegler, Mark (1981). The Doctor-Patient Encounter and Its Relation to Theories of Health and Disease. In: A.L. Caplan et al. (Eds.), *Concepts of Health and Disease: Interdisciplinary Perspectives* (pp. 627-644). Massachusetts: Addison-Wesley Publishing Company.

Sigerist, Henry E. (1943). *Civilization and Disease.* Chicago: University of Chicago Press.

Smith-Rosenberg, Carroll, and Rosenberg, Charles (1981). The Female Animal: Medical and Biological Views of Woman and Her Role in Nineteenth-Century America. In: A.L. Caplan et al. (Eds.), *Concepts of Health and Disease: An Interdisciplinary Perspective* (pp. 281-303). Massachusetts: Addison-Wesley.

Sontag, Susan (1979). *Disease as Metaphor.* London: Allen Lane.

Sontag, Susan (1990). *Illness as Metaphor; And AIDS and Its Metaphors.* New York: Doubleday.

Sorenson, J.R. (1974). Biomedical Innovation, Uncertainty, and Doctor-Patient Interaction. *Journal of Health and Social Behavior 15*, 366-374.

Spector, Rachel E. (1996). *Cultural Diversity in Health and Disease*, 4th ed. Connecticut: Appleton and Lange.

Stapleton, Julia (ed.) (1995). *Group Rights: Perspectives Since 1900.* England: Thoemmes Press.

Stevens, M.L. Tina (2000). *Bioethics in America: Origins and Cultural Politics.* Maryland: Johns Hopkins University Press.

Stevenson, Charles L. (1944). *Ethics and Language.* Connecticiut: Yale University Press.

Stokinger, H.D. et al. (1973). Hypersusceptibility and Genetic Problems in Occupational Medicine: A Consensus Report. *Journal of Occupational Medicine, 15*, 564-573.

Sündstöm, Per (1987). *Icons of Disease: A Philosophical Inquiry Into the Semantics, Phenomenology, and Ontology of the Clinical Conceptions of Disease.* Stockholm: Linkoping.

Sussman, Norman, and Hyler, Steven E. (1986). Factitious Disorders. *Harrison's Textbook of Medicine* 8th ed. (pp. 1242-1247). New York: MacMillan.

Sydenham, Thomas (1981 [1676]) Preface to the Third Edition. In: A.L. Caplan et al., eds. *Concepts of Health and Disease: Interdisciplinary Perspectives* (pp. 145-155). Massachusetts: Addison-Wesley, Publishing Company.

Szasz, Thomas (1961). *Myth of Mental Illness*. New York: Harper-Hoeber.

Szasz, Thomas (1973). *The Second Sin*. London: Routledge and Kegan Paul.

Tavris, Carol (1992). *Mismeasure of Woman*. New York: Simon and Schuster.

Temkin, Oswei (1981). The Scientific Approach to Disease: Specific Entity and Individual Sickness. In: A.L. Caplan et al. (Eds.), *Concepts of Health and Disease: Interdisciplinary Perspectives* (pp. 247-263). Massachusetts: Addison-Wesley Publishing Company.

Thompson, Becky W. (1994). *A Hunger So Wide and So Deep: American Women Speak Out On Eating Problems*. Minnesota: University of Minnesota Press.

Tobias, Andrew (1973). *The Best Little Boy in the World*. New York: Ballantine.

Toews, John E. (1987). Intellectual History After the Linguistic Turn: The Autonomy of Meaning and the Irreducibility of Experience. *American Historical Review, 92*, 879-907.

Tong, Rosemarie (1997). *Feminist Approaches to Bioethics: Theoretical Reflection and Practical Applications*. Colorado: Westview Press.

Tong, Rosemarie, with G. Anderson and A. Santos (Eds.). (2000). *Globalizing Feminist Bioethics: Cross-cultural Perspectives*. Colorado: Westview Press.

Toombs, S. Kay (1992). *The Meaning of Illness: A Phenomenological Account of the Different Perspectives of Physician and Patient*. Kluwer Academic Publishers, Dordrecht, Holland.

Torkelson, Jean (2000). Faith Healing of Children Under Fire. *Denver Rocky Mountain News, April 14,* 5A-6A

Toulmin, Stephen (1960). *Foresight and Understanding*, New York: Harper and Row Publishers.

Treichler, Paula A. (1999). *How to Have Theory in an Epidemic: Cultural Chronicles of AIDS*. North Carolina: Duke University Press.

Trotter, Thomas (1804). *An Essay, Medical, Philosophical, and Chemical, on Drunkenness and Its Effect on the Human Body*. London: Logman, Reese, and Orme.

Tuana, Nancy (1988). The Weaker Seed: The Sexist Bias of Reproductive Theory. *Hypatia, 3, 1,* 35-59.

Varmus, H. (1989). Naming the AIDS Virus. In: E.T Juengst and B.A. Koenig (Eds.), *Meaning of AIDS: Perspectives From the Humanities* (pp. 3-11). New York: Praeger Scientific Publishers.

Veatch, Robert M. (1992). *Cross-Cultural Perspectives on Medical Ethics*. Boston: Jones and Bartlett.

Veatch, Robert M. (1995). Bioethics and the Philosophy of Science. *The Journal of Medicine and Philosophy 20*, 227-231.

Veatch, Robert M. (1997). *Medical Ethics*, 2nd ed. Boston: Jones and Bartlett Publishers, Inc.

Virchow, Rudolf (1981 [1858]). Three Selections from Rudolf Virchow. In: A.L. Caplan et al. (Eds.), *Concepts of Health and Disease: Interdisciplinary Perspectives* (pp. 187-195). Massachusetts: Addison-Wesley Publishing Company.

Virchow, Rudolf (1981 [1895]). One Hundred Years of General Pathology. In H.T. Engelhardt, Jr. et al. (Eds.), *Concepts of Health and Disease: Interdisciplinary Perspectives* (pp. 190-195). Massachusetts: Addison-Wesley Publishing Company.

Vogel, Morris J., and Rosenberg, Charles E. (Eds.) (1979). *The Therapeutic Revolution: Essays in the Social History of American Medicine*. Pennsylvania: University of Philadelphia Press.

von Wright, George Henrik (1963). *The Varieties of Goodness*. New York: Humanities Press.

Waisberg, J., and Paige, P. (1988). Gender Role Conformity and Perception of Mental Illness. *Women's Health*   14, 3-16.

Wallis, Claudia (1985). Aids: A Growing Threat. *Time, 126 , August 12*, 41.

Walters, LeRoy (1997). Reproductive Technologies and Genetics. In: R.M. Veatch (Ed.), *Medical Ethics* (pp. 209-238). Massachusetts: Jones and Bartlett Publishers.

Warnock, Geoffrey J. (1971). *The Object of Morality*. London: Methuen and Co. Ltd.

Wartofsky, Marx (1975). Organs, Organisms, and Disease. In: H.T. Engelhardt, Jr. and S,F, Spicker (Eds.), *Evaluation and Explanation in the Biomedical Sciences* (pp. 67-83). Holland: D. Reidel Publishing Company.

Wartofsky, Marx (1976). The Mind's Eye and the Hand's Brain: Toward an Historical Epistemology of Medicine. In : H.T. Engelhardt, Jr. and D. Callahan (Eds.), *Science, Ethics, and Medicine* (pp. 167-194). New York: The Hastings Center.

Wartofsky, Marx (1977). How to Begin Again: Medical Therapies for the Philosophy of Science. In: F. Suppe and P.D. Asquith (Eds.), *PSA 1976*, Vol. Two (pp. 4-108). Michigan: Philosophy of Science Association.

Wartofsky, Marx (1992). The Social Presuppositions of Medical Knowledge. In: J.L. Peset and D. Garcia (Eds.), *The Ethics of Diagnosis* (pp. 131-151). Holland: Kluwer Academic Publishers.

Watson, James (1968). *The Double Helix*. New York: Atheneum.

Watson, James D., and Crick, Francis H. (1953). Molecular Structure of Nucleic Acids. *Nature, 171,* 737.

Weiss, Rick (2001). Genome Puzzle Yields Big Picture. *The Sunday Denver Post. February 11*, A1, 18A.

Weissman, Myrna M., and Olfson, Mark (1995). Depression in Women: Implications for Health Care Research. *Science, 269, August 11*, 799-801.

Whitbeck, Caroline (1981). A Theory of Health. In: A.L. Caplan et al. (Eds.), *Concepts of Health and Disease: Interdisciplinary Perspectives* (pp. 611-626). Massachusetts: Addison-Wesley Publishing Company.

Williams, G.C. and Neese, R.M. (1991). The Dawn of Darwinian Medicine. *Quarterly Review of Biology, 66, 1,* X

Wofsy, C.B. et al. (1986). Isolation of AIDS-Associated Retrovirus from Genital Secretions of Women with Antibodies to the Virus. *Lancet, 8, 1*, 527-529.

Wolf, Stewart (1961). Disease as a Way of Life: Neural Integration in Systematic Pathology. *Perspectives in Biology and Medicine, 4*, 288-305.

Wolf, Susan (1996). *Feminism and Bioethics: Beyond Reproduction*. New York: Oxford University Press.

World Health Association (2003). Nutrition, Health, and Human Rights. *World Health Association*, March 11, March 11 <www.who.int/nut/nutrition.htm>

Wulff, Henrik (1981a). How to Make the Best Decisions. *Medical Decision-making, 1*, 277-283.

Wulff, Henrik R. (1981b). *Rational Diagnosis and Treatment: An Introduction to Clinical Decision-Making*, 2nd ed., Oxford: Blackwell Scientific Publications.

Wulff, Henrik R. (1984). The Causal Basis of the Current Disease Classification. In: I. Lindahl and L. Nordenfelt (Eds.), *Health, Disease, and Causal Explanations in Medicine* (pp. 169-177). Holland: D. Reidel Publishing Company.

Wulff, Henrik R. et al. (1986). *Philosophy of Medicine: An Introduction*. Oxford: Blackwell Scientific Publications.

Young, Robert M. (1970). *Mind, Brain, and Adaptation in the Nineteenth Century*. Oxford: Clarendon Press.

INDEX

186   INDEX

Philosophy and Medicine

1. H. Tristram Engelhardt, Jr. and S.F. Spicker (eds.): *Evaluation and Explanation in the Biomedical Sciences.* 1975   ISBN 90-277-0553-4
2. S.F. Spicker and H. Tristram Engelhardt, Jr. (eds.): *Philosophical Dimensions of the Neuro-Medical Sciences.* 1976   ISBN 90-277-0672-7
3. S.F. Spicker and H. Tristram Engelhardt, Jr. (eds.): *Philosophical Medical Ethics.* Its Nature and Significance. 1977   ISBN 90-277-0772-3
4. H. Tristram Engelhardt, Jr. and S.F. Spicker (eds.): *Mental Health.* Philosophical Perspectives. 1978   ISBN 90-277-0828-2
5. B.A. Brody and H. Tristram Engelhardt, Jr. (eds.): *Mental Illness.* Law and Public Policy. 1980   ISBN 90-277-1057-0
6. H. Tristram Engelhardt, Jr., S.F. Spicker and B. Towers (eds.): *Clinical Judgment.* A Critical Appraisal. 1979   ISBN 90-277-0952-1
7. S.F. Spicker (ed.): *Organism, Medicine, and Metaphysics.* Essays in Honor of Hans Jonas on His 75th Birthday. 1978   ISBN 90-277-0823-1
8. E.E. Shelp (ed.): *Justice and Health Care.* 1981
                                    ISBN 90-277-1207-7; Pb 90-277-1251-4
9. S.F. Spicker, J.M. Healey, Jr. and H. Tristram Engelhardt, Jr. (eds.): *The Law-Medicine Relation.* A Philosophical Exploration. 1981   ISBN 90-277-1217-4
10. W.B. Bondeson, H. Tristram Engelhardt, Jr., S.F. Spicker and J.M. White, Jr. (eds.): *New Knowledge in the Biomedical Sciences.* Some Moral Implications of Its Acquisition, Possession, and Use. 1982   ISBN 90-277-1319-7
11. E.E. Shelp (ed.): *Beneficence and Health Care.* 1982   ISBN 90-277-1377-4
12. G.J. Agich (ed.): *Responsibility in Health Care.* 1982   ISBN 90-277-1417-7
13. W.B. Bondeson, H. Tristram Engelhardt, Jr., S.F. Spicker and D.H. Winship: *Abortion and the Status of the Fetus.* 2nd printing, 1984   ISBN 90-277-1493-2
14. E.E. Shelp (ed.): *The Clinical Encounter.* The Moral Fabric of the Patient-Physician Relationship. 1983   ISBN 90-277-1593-9
15. L. Kopelman and J.C. Moskop (eds.): *Ethics and Mental Retardation.* 1984
                                                 ISBN 90-277-1630-7
16. L. Nordenfelt and B.I.B. Lindahl (eds.): *Health, Disease, and Causal Explanations in Medicine.* 1984   ISBN 90-277-1660-9
17. E.E. Shelp (ed.): *Virtue and Medicine.* Explorations in the Character of Medicine. 1985   ISBN 90-277-1808-3
18. P. Carrick: *Medical Ethics in Antiquity.* Philosophical Perspectives on Abortion and Euthanasia. 1985   ISBN 90-277-1825-3; Pb 90-277-1915-2
19. J.C. Moskop and L. Kopelman (eds.): *Ethics and Critical Care Medicine.* 1985
                                                 ISBN 90-277-1820-2
20. E.E. Shelp (ed.): *Theology and Bioethics.* Exploring the Foundations and Frontiers. 1985   ISBN 90-277-1857-1

Philosophy and Medicine

21. G.J. Agich and C.E. Begley (eds.): *The Price of Health*. 1986
                                                    ISBN 90-277-2285-4
22. E.E. Shelp (ed.): *Sexuality and Medicine*. Vol. I: Conceptual Roots. 1987
                                          ISBN 90-277-2290-0; Pb 90-277-2386-9
23. E.E. Shelp (ed.): *Sexuality and Medicine*. Vol. II: Ethical Viewpoints in Transition.
     1987   ISBN 1-55608-013-1; Pb 1-55608-016-6
24. R.C. McMillan, H. Tristram Engelhardt, Jr., and S.F. Spicker (eds.): *Euthanasia
     and the Newborn*. Conflicts Regarding Saving Lives. 1987
                                          ISBN 90-277-2299-4; Pb 1-55608-039-5
25. S.F. Spicker, S.R. Ingman and I.R. Lawson (eds.): *Ethical Dimensions of Geriatric
     Care*. Value Conflicts for the 21th Century. 1987   ISBN 1-55608-027-1
26. L. Nordenfelt: *On the Nature of Health*. An Action-Theoretic Approach. 2nd,
     rev. ed. 1995   SBN 0-7923-3369-1; Pb 0-7923-3470-1
27. S.F. Spicker, W.B. Bondeson and H. Tristram Engelhardt, Jr. (eds.): *The Contra-
     ceptive Ethos*. Reproductive Rights and Responsibilities. 1987
                                                    ISBN 1-55608-035-2
28. S.F. Spicker, I. Alon, A. de Vries and H. Tristram Engelhardt, Jr. (eds.): *The Use
     of Human Beings in Research*. With Special Reference to Clinical Trials. 1988
                                                    ISBN 1-55608-043-3
29. N.M.P. King, L.R. Churchill and A.W. Cross (eds.): *The Physician as Captain of
     the Ship*. A Critical Reappraisal. 1988   ISBN 1-55608-044-1
30. H.-M. Sass and R.U. Massey (eds.): *Health Care Systems*. Moral Conflicts in
     European and American Public Policy. 1988   ISBN 1-55608-045-X
31. R.M. Zaner (ed.): *Death: Beyond Whole-Brain Criteria*. 1988
                                                    ISBN 1-55608-053-0
32. B.A. Brody (ed.): *Moral Theory and Moral Judgments in Medical Ethics*. 1988
                                                    ISBN 1-55608-060-3
33. L.M. Kopelman and J.C. Moskop (eds.): *Children and Health Care*. Moral and
     Social Issues. 1989   ISBN 1-55608-078-6
34. E.D. Pellegrino, J.P. Langan and J. Collins Harvey (eds.): *Catholic Perspectives
     on Medical Morals*. Foundational Issues. 1989   ISBN 1-55608-083-2
35. B.A. Brody (ed.): *Suicide and Euthanasia*. Historical and Contemporary Themes.
     1989   ISBN 0-7923-0106-4
36. H.A.M.J. ten Have, G.K. Kimsma and S.F. Spicker (eds.): *The Growth of Medical
     Knowledge*. 1990   ISBN 0-7923-0736-4
37. I. Löwy (ed.): *The Polish School of Philosophy of Medicine*. From Tytus
     Chałubiński (1820–1889) to Ludwik Fleck (1896–1961). 1990
                                                    ISBN 0-7923-0958-8
38. T.J. Bole III and W.B. Bondeson: *Rights to Health Care*. 1991
                                                    ISBN 0-7923-1137-X

Philosophy and Medicine

Philosophy and Medicine

Philosophy and Medicine

KLUWER ACADEMIC PUBLISHERS – DORDRECHT / BOSTON / LONDON